Praise for *Delusions of Gend*

'Impeccably researched and bitingly funny … both sexes should rejoice at [this] vitriolic attack on … sexism masquerading as psychology.' *Evening Standard*

'Bold … Timely and provocative … [Fine's] well-stocked armoury … includes extensive research, sharp wit and a probing intelligence, and refuses to be satisfied with the delusional myth-making that often passes for popular science.' *Metro*

'Fine writes with bravura. She takes no hostages. She rejoices in demystifying the compellingly seductive false colour images provided by the MRI scanners … a book that sparkles with wit, which is easy to read but underpinned by substantial scholarship and a formidable 100-page bibliography … every page of Fine's brilliant, spiky book reminds us that science is part of culture and that the struggle against sexism in the neurosciences and the struggle against sexism in society are intimately linked.' **Hilary Rose**, *Times Higher Education Supplement*

'Fine invites her readers into a passionate, insightful and often funny discussion about how gender identity is all in the mind, not the brain.' *Globe & Mail*, **Canada**

'As Fine argues in this forceful, funny new book, the notion that gender accounts for differences in minds and behavior through some biological, brain-based process is an idea as popular as it is unproven.' *Boston Globe*

'An irreverent and important book' *Washington Post*

'Read this book and see how complex and fascinating the whole issue is.' *New York Times*

'A timely warning against taking too seriously the deluge of books and articles that would have us believe that men are biologically advantaged when it comes to mathematics, racing driving or map reading – and that women are naturally more intuitive and nurturing, so better at childcare and multitasking.' *Guardian*

'Dr Fine is a brilliant tour guide – making light, fun and engaging work of the research. By debunking the rubbish, this book opens up possibilities for a (slightly) clearer vision of the future. Not to be missed.' **www.fat-quarter.co.uk**

'Men may be from Mars and women from Venus but if you put blokes and sheilas on each other's planet they will work out how to manage … An excellent book that puts the old nature-or-nurture debate in the context of the new science on the way our brains work.' *The Australian*

CORDELIA FINE

The Real Science
Behind Sex Differences

DELUSIONS
OF GENDER

ICON BOOKS

First published in the UK in 2010 by Icon Books

This edition published in the UK in 2011 by
Icon Books Ltd, Omnibus Business Centre,
39–41 North Road, London N7 9DP
email: info@iconbooks.co.uk
www.iconbooks.co.uk

Reprinted 2011

First published in the USA in 2010 by
W.W. Norton & Company, Inc., New York

Sold in the UK, Europe, South Africa and Asia
by Faber & Faber Ltd, Bloomsbury House,
74–77 Great Russell Street,
London WC1B 3DA or their agents

Distributed in the UK, Europe, South Africa and Asia
by TBS Ltd, TBS Distribution Centre, Colchester Road,
Frating Green, Colchester CO7 7DW

Published in Australia in 2011
by Allen & Unwin Pty Ltd, PO Box 8500,
83 Alexander Street, Crows Nest, NSW 2065

ISBN: 978-184831-220-3

Printed and bound in the UK by

Clays Ltd, St Ives plc

For my mother

ABOUT THE AUTHOR

Cordelia Fine is a Research Associate at the Centre for Agency, Values and Ethics at Macquarie University, Australia, and an Honorary Research Fellow at the Department of Psychological Sciences at the University of Melbourne, Australia. Her previous book, *A Mind of Its Own* (Icon, 2006) was hugely acclaimed and she was called 'a science writer to watch' by *Metro*.

CONTENTS

PART 2

NEUROSEXISM

PART 3

RECYCLING GENDER

DELUSIONS OF GENDER

Of all difficulties which impede the progress of thought, and the formation of well-grounded opinions on life and social arrangements, the greatest is now the unspeakable ignorance and inattention of mankind in respect to the influences which form human character. Whatever any portion of the human species now are, or seem to be, such, it is supposed, they have a natural tendency to be: even when the most elementary knowledge of the circumstances in which they have been placed, clearly points out the causes that made them what they are.

—John Stuart Mill, *The Subjection of Women* (1869)

INTRODUCTION

Meet Evan.

When his wife, Jane, is upset, he sits with her on the couch, reading a magazine or book 'to distract himself from his own discomfort' while he cradles Jane with the other arm. After a few years working on this issue, Evan gradually comes to be able to offer comfort in a more conventional way. The politically correct and/or scientifically uninformed among you may be wondering about the cause of Evan's peculiar behaviour. Does he secretly find Jane deeply unattractive? Is he in the slow process of recovery from some deeply traumatic incident? Was he raised by wolves until the age of thirteen? Not at all. He's just a regular guy, with a regular guy-brain that's wired all wrong for empathy. That a simple act of comfort is not part of Evan's behavioural repertoire is the fault of the neurons dealt him by nature: neurons that endure a devastating 'testosterone marination'; neurons that are lacking the same 'innate ability to read faces and tone of voice for emotional nuance' as women's; neurons, in a word, that are male.[1]

Evan is just one of several curious characters who populate Louann Brizendine's *New York Times* best seller, *The Female Brain*. In her depiction, men's empathising skills resemble those of the hapless tourist attempting to decipher a foreign menu and are sharply contrasted with the cool proficiency of females' achievements in this domain. Take Sarah, for example. Sarah can 'identify and anticipate what [her husband] is feeling – often before he is conscious of it himself.' Like the magician who knows that you'll pick the seven of diamonds even before it's left the pack, Sarah can amaze her husband at whim, thanks to her lucky knack of knowing what he's feeling before he feels it. (*Ta-DA! Is* this *your emotion?*) And no, Sarah is not a fairground psychic. She is simply a woman who enjoys the extraordinary gift of mind reading that, apparently, is bestowed on all owners of a female brain:

> Maneuvring like an F-15, Sarah's female brain is a high-performance emotion machine – geared to tracking, moment by moment, the non-verbal signals of the innermost feelings of others.[2]

Just what is it that makes the female brain so well suited to stalking people's private feelings as though they were terrified prey? Why, you are asking, are male neurons not capable of such miracles – better placed instead to navigate the masculine worlds of science and maths? Whatever the answer du jour – whether it's the foetal testosterone that ravages the male neural circuits, the oversized female corpus callosum, the efficiently specialised organisation of the male brain, the primitively subcortical emotion circuits of boys, or the underendowment of visuospatial processing white matter in the female brain – the underlying message is the same. Male and female brains are different in ways that matter.

Having marital problems, for instance? Turn to *What Could He Be Thinking?* by 'educator, therapist, corporate consultant, and . . . *New York Times* bestselling author'[3] Michael Gurian, and you will discover the epiphany the author experienced with his wife, Gail, on seeing MRI (magnetic resonance imaging) and PET (positron emission tomography) scans of male and female brains:

> I said, 'We thought we knew a lot about each other, but maybe we haven't known enough.' Gail said, 'There really is such a thing as a "male" brain. It's hard to argue with an MRI.' We

> realized that our communication, our support of each other,
> and our understanding of our relationship were just beginning,
> after six years of marriage.

The information from those scans, says Gurian, was 'marriage saving.'[4]

Nor are spouses the only ones who, it is now claimed, can be better understood with the benefit of a little background in brain science. The blurb of the influential book *Why Gender Matters* by physician Leonard Sax, founder and executive director of the National Association for Single Sex Public Education (NASSPE), promises to show readers how to 'recognize and understand . . . hardwired differences [between the sexes] to help every girl and every boy reach their fullest potential.'[5] Likewise, parents and teachers are informed in a recent Gurian Institute book that 'Researchers [using MRI] have literally seen what we have always known. There are fundamental gender differences and they start in the very structure of the human brain.'[6] Thus, Gurian suggests that 'to walk into a classroom or home without knowledge of both how the brain works and how the male and female brains learn differently is to be many steps behind where we can and should be as teachers, parents, and caregivers of children.'[7]

Even CEOs can, it is said, benefit from a greater understanding of sex differences in the brain. The recent book *Leadership and the Sexes* 'links the actual science of male/female brain differences to every aspect of business' and 'presents brain science tools with which readers can look into the brains of men and women to understand themselves and one another.' According to the jacket blurb, the 'gender science' in the book 'has been used successfully by such diverse corporations as IBM, Nissan, Proctor [*sic*] & Gamble, Deloitte & Touche, PriceWaterhouseCoopers, Brooks Sports, and many others.'[8]

Is it realistic, you will begin to wonder, to expect two kinds of people, with such different brains, to ever have similar values, abilities, achievements, lives? If it's our differently wired brains that make us different, maybe we can sit back and relax. If you want the answer to persisting gender inequalities, stop peering suspiciously at society and take a look right over here, please, at this brain scan.

If only it were that simple.

About 200 years ago, the English clergyman Thomas Gisborne wrote a book that despite its, to my mind, rather unappealing title – *An Enquiry into the Duties of the Female Sex* – became an eighteenth-century best seller. In it, Gisborne neatly set out the different mental abilities required to fulfil male versus female roles:

> The science of legislation, of jurisprudence, of political economy; the conduct of government in all its executive functions; the abstruse researches of erudition . . . the knowledge indispensable in the wide field of commercial enterprise . . . these, and other studies, pursuits, and occupations, assigned chiefly or entirely to men, demand the efforts of a mind endued with the powers of close and comprehensive reasoning, and of intense and continued application.[9]

It was only natural, the author argued, that these qualities should be 'impart[ed] . . . to the female mind with a more sparing hand' because women have less need of such talents in the discharge of their duties. Women are not inferior, you understand, simply *different*. After all, when it comes to performance in the feminine sphere 'the superiority of the female mind is unrivalled', enjoying 'powers adapted to unbend the brow of the learned, to refresh the over-laboured faculties of the wise, and to diffuse, throughout the family circle, the enlivening and endearing smile of cheerfulness'.[10] What awfully good luck that these womanly talents should coincide so happily with the duties of the female sex.

Fast-forward 200 years, turn to the opening page of *The Essential Difference*, a highly influential twenty-first-century book about the psychology of men and women, and there you will find Cambridge University psychologist Simon Baron-Cohen expressing much the same idea: *'The female brain is predominantly hard-wired for empathy. The male brain is predominantly hard-wired for understanding and building systems.'*[11] Just like Gisborne, Baron-Cohen thinks that it is those with the 'male brain' who make the best scientists, engineers, bankers and lawyers, thanks to their capacity to focus in on different aspects of a system (be it a biological, physical, financial or legal system), and their drive to understand how it works. And the soothing reassurance that women, too, have their own special talents remains present and correct. In what has been described as a 'masterpiece of condescension',[12] Baron-Cohen explains that the female brain's propensity for understanding others' thoughts and feelings, and responding to them sympathetically, ideally suits it to occupations that professionalise women's traditional caring roles: 'People with the female brain make the most wonderful counsellors, primary-school teachers, nurses, carers, therapists, social workers, mediators, group facilitators or personnel staff.'[13] Philosopher Neil Levy's neat summary of Baron-Cohen's thesis – that 'on average, women's intelligence is best employed in putting people at their ease, while the men get on with understanding the world and building and repairing the things we need in it'[14] – can't help but bring to mind Gisborne's eighteenth-century wife, busily unbending the brow of her learned husband.

Baron-Cohen does, it must be said, take great pains to point out that not all women have a female, empathising brain, nor all men a male, systemising one. However, this concession does not set him apart from traditional views of sex differences quite as much as he might think. As long ago as 1705, the philosopher Mary Astell observed that women who made great achievements in male domains were said by men to have *'acted above their Sex. By which one must suppose they wou'd have their Readers understand, That* they were not Women who did those Great Actions, but that they were Men in Petticoats!'[15] Likewise, a few centuries later intellectually talented women were 'said to possess "masculine minds".'[16] As one writer opined in the *Quarterly Journal of Science*:

> The *savante* – the woman of science – like the female athlete, is simply an anomaly, an exceptional being, holding a position more or less intermediate between the two sexes. In one case the brain, as in the other the muscular system, has undergone an abnormal development.[17]

Baron-Cohen, of course, does not describe as 'abnormal' the woman who reports a greater tendency to systemise. But certainly there is an incongruous feel to the idea of a male brain in the body of a woman, or a female brain housed in the skull of a man.

The sheer stability and staying power of the idea that male and female psychologies are inherently different can't help but impress. Are there, in truth, psychological differences hardwired into the brains of the sexes that explain why, even in the most egalitarian of twenty-first-century societies, women and men's lives still follow noticeably different paths?

For many people, the experience of becoming a parent quickly abolishes any preconceptions that boys and girls are born more or less the same. When the gender scholar Michael Kimmel became a father, he reports that an old friend cackled to him, 'Now you'll see it's all biological!'[18] And what could be more compelling proof of this, as a parent, than to see your own offspring defy your well-meaning attempts at gender-neutral parenting? This is a common experience, discovered sociologist Emily Kane. Many parents of preschoolers – particularly the white, middle- and upper-middle-class ones – came to the conclusion that differences between boys and girls were biological by process of elimination. Believing that they practised gender-neutral parenting, the 'biology as fallback' position, as Kane calls it, was the only one left remaining to them.[19]

Some commentators, casting their eye over society at large,

find themselves falling back on biology in much the same way. In her recent book *The Sexual Paradox*, journalist and psychologist Susan Pinker tackles the question of why 'gifted, talented women with the most choices and freedoms don't seem to be choosing the same paths, in the same numbers, as the men around them. Even with barriers stripped away, they don't behave like male clones.' Considering this, to some, unexpected outcome, Pinker wonders 'whether biology is, well, if not destiny exactly, then a profound and meaningful departure point for a discussion about sex differences.'[20] The gender gap, she suggests, has in part 'neurological or hormonal roots'.[21] As the barriers of a sexist society continue to fall, there seem to be fewer and fewer social scapegoats to call on to explain continuing gender inequalities and work segregation. When we can't pin the blame on outside forces, all eyes swivel to the internal – the differences in the structure or functioning of female and male brains. Wired differently from men, many women choose to reject what Pinker calls the 'vanilla' male model of life – in which career takes priority over family – and have different interests.

The fallback conclusion that there must be hardwired psychological differences between the sexes also appears to enjoy impressive scientific support. First, there is the surge of foetal testosterone that takes place during the gestation of male, but not female, babies. As *Brain Sex* authors Anne Moir and David Jessel describe this momentous event:

> [At] six or seven weeks after conception . . . the unborn baby 'makes up its mind', and the brain begins to take on a male or a female pattern. What happens, at that critical stage in the darkness of the womb, will determine the structure and organisation of the brain: and that, in turn, will decide the very nature of the mind.[22]

Like other popular writers, Moir and Jessel leave us in little danger of underestimating the psychological significance of what goes on 'in the darkness of the womb'. While Louann Brizendine is content to merely state that the effect of prenatal testosterone on the brain 'defines our innate biological destiny',[23] Moir and Jessel are openly gleeful about the situation. '[Infants] have, quite literally, made up their minds in the womb, safe from the legions of social engineers who impatiently await them.'[24]

Then, there are the differences between male and female brains. Rapid progress in neuroimaging technology enables neuroscientists to see, in ever-increasing detail, sex differences in brain structure and function. Our brains are different, so surely our minds are too? For example, in a *New York Times Magazine* feature on the so-called opt-out revolution (that is, women who give up their careers to take up traditional roles as stay-at-home mothers) one interviewee told journalist Lisa Belkin that '"[i]t's all in the M.R.I.," . . . [referring to] studies that show the brains of men and women "light up" differently when they think or feel. And those different brains, she argues, inevitably make different choices.'[25] The neuroscientific discoveries we read about in magazines, newspaper articles, books and sometimes even journals tell a tale of two brains – essentially different – that create timeless and immutable psychological differences between the sexes. It's a compelling story that offers a neat, satisfying explanation, and justification, of the gender status quo.[26]

We have been here before, so many times.

In the seventeenth century, women were severely disadvantaged educationally; for example, in their political development they were hindered 'through their lack of formal education in political rhetoric, their official exclusion from citizenship and government, the perception that women ought not to be involved in political affairs, and the view that it was immodest for a woman to write at all.'[27] Yet despite such – to our modern eyes – obvious impediments to women's intellectual development, they were widely assumed to be naturally inferior by many. While, in retrospect, it might seem to go without saying that men's apparently superior intellect and achievements might lie in sources other than natural neural endowments, at the time it *did* need saying. As one seventeenth-century feminist put it: 'For a Man ought no more to value himself upon being Wiser than a Woman, if he owe his Advantage to a better Education, and greater means of Information, then he ought to boast of his Courage, for beating a Man, when his Hands were bound'.[28]

In the eighteenth century, as we've seen, Thomas Gisborne felt no need to consider an alternative explanation of his observations of sex differences within society. As the writer Joan Smith has pointed out:

> [V]ery few women, growing up in England in the late eighteenth century, would have understood the principles of jurisprudence or navigation, but that is solely because they were denied access to them. Obvious as this is to a modern observer, the hundreds of thousands of readers who bought his books accepted his argument at face value because it fitted in with their prejudices.[29]

And in the late nineteenth and early twentieth centuries, women still did not have equal access to higher education. And yet, '[w]omen', declared the well-known psychologist Edward Thorndike, 'may and doubtless will be scientists and engineers, but the Joseph Henry, the Rowland, and the Edison of the future, will be men'. This confident proclamation, made at a time when women were not granted full membership to, for example, Harvard, Cambridge or Oxford University seems – I don't know – a bit premature? And, given that at the time women couldn't vote, was it not also a little rash for Thorndike to claim with such confidence that 'even should all women vote, they would play a small part in the Senate'?[30] In retrospect, the constraints on women are perfectly obvious. *Hey, Professor Thorndike,* we might think to ourselves, *ever think about letting women into the Royal Society, or maybe offering them a little civil entitlement known as the vote, before casting judgement on their limitations in science and politics?* Yet to many of those who were there at the time, the slope of the playing field was imperceptible. Thus philosopher John Stuart Mill's denial in 1869 that 'any one knows, or can know, the nature of the two sexes, as long as they have only been seen in their present relation to one another'[31] was revolutionary, and derided. Decades later it was still with only the utmost tentativeness that the early-twentieth-century researcher of 'eminence', Cora Castle, asked, 'Has innate inferiority been the reason for the small number of eminent women, or has civilisation never yet allowed them an opportunity to develop their innate powers and possibilities?'[32]

There is also nothing new about looking to the brain to explain and justify the gender status quo. In the seventeenth century, the French philosopher Nicolas Malebranche declared women 'incapable of penetrating to truths that are slightly difficult to discover', claiming that '[e]verything abstract is incomprehensible to them.' The neurological explanation for this, he proposed, lay in the 'delicacy of the brain fibers'.[33] Presumably, one abstract thought too many and – *ping!* – those fibres snap. Over the intervening centuries, the neurological explanations behind men and women's different roles, occupations and achievements have been overhauled again and again, as neuroscientific techniques and understanding have become ever more sophisticated. Early brain scientists, using the cutting-edge techniques of the time, busily filled empty skulls with pearl barley, carefully categorised head shape using tape measures and devoted large portions of careers to the weighing of brains.[34] Infamously, they proposed that women's intellectual inferiority stemmed from their smaller and lighter brains, a phenomenon that came to be widely known among the Victorian public as 'the missing five ounces of the female brain.'[35] The hypothesis, widely believed, that this sex difference in the brain was of profound psychological significance was championed by Paul Broca, one of the most eminent scientists of the time. Only when it became inescapably clear that brain weight did not correlate with intelligence did brain scientists acknowledge that men's larger brains might merely reflect their larger bodies. This inspired a search for a measure of relative, rather than absolute, brain weight that would leave the absolutely bigger-brained sex ahead. As historian of science Cynthia Russett reports:

> Many ratios were tried – of brain weight to height, to body weight, to muscular mass, to the size of the heart, even (one begins to sense desperation) to some one bone, such as the femur.[36]

These days, we have rather more of an inkling of the complexity of the brain. It's undeniable that by moving into the realm of the brain itself, rather than its outer casing, scientific advance was made. It was certainly an important moment when a forward-thinking nineteenth-century scientist, fingering his tape measure with the tense distraction of one who suspects that his analysis has left certain important details unpenetrated, said thoughtfully, 'Pass me that brain and those scales, will you?' But even the untrained twenty-first-century layperson can see that this brought scientists only a little closer to understanding the mystery of how brain cells create the engine of the mind, and can sense the unfortunate hastiness of the conclusion that women's cognitive inferiority to men could be weighed in ounces.

It may seem like the same sort of prejudice couldn't possibly creep into the contemporary debate because now we are all so enlightened; perhaps even . . . *over*enlightened? Writers who argue that there are hardwired differences between the sexes that account for the gender status quo often like to position themselves as courageous knights of truth, who brave the stifling ideology of political correctness. Yet claims of 'essential differences' between the two sexes simply reflect – and give scientific authority to – what I suspect is really a majority opinion.[37] If history tells us anything, it is to take a second, closer look at our society and our science. This is the aim of *Delusions of Gender*.

At the core of the first part of this book, '"Half-Changed World", Half-Changed Minds', is the critical idea that the psyche is 'not a discrete entity packed in the brain. Rather, it is a structure of psychological processes that are shaped by and thus closely attuned to the culture that surrounds them.'[38] We tend not to think about ourselves this way, and it's easy to underestimate the impact of what is *outside* the mind on what takes place inside. When we confidently compare the 'female mind' and the 'male mind', we think of something stable inside the head of the person, the product of a 'female' or 'male' brain. But such a tidily isolated data processor is not the mind that social and cultural psychologists are getting to know with ever more intimacy. As Harvard University psychologist Mahzarin Banaji puts it, there is no 'bright line separating self from culture', and the culture in which we develop and function enjoys a 'deep reach' into our minds.[39] It's for this reason that we can't understand gender differences in female and male minds – the minds that are the source of our thoughts, feelings, abilities, motivations, and behaviour – without understanding how psychologically permeable is the skull that separates the mind from the sociocultural context in which it operates. When the environment makes gender salient, there is a ripple effect on the mind. We start to think of ourselves in terms of our gender, and stereotypes and social expectations become more prominent in the mind. This can change self-perception, alter interests, debilitate or enhance ability, and trigger unintentional discrimination. In other words, the social context influences who you are, how you think and what you do. And these thoughts, attitudes and behaviours of yours, in turn, become part of the social context. It's intimate. It's messy. And it demands a different way of thinking about gender.

Then, there's the less subtle, consciously performed discrimination against women, the wide-ranging forms of exclusion, the harassment and the various injustices both at work and home. These stem from not-all-that-old, and still powerful, ideas about men and women's proper roles and places in the world. By the end of the first part of the book, one can't help but wonder if we have stumbled on the twenty-first-century blind-spot. As University of California–Irvine professor of mathematics Alice Silverberg commented:

> When I was a student, women in the generation above me told horror stories about discrimination, and added 'But everything has changed. That will never happen to you.' I'm told that this was said even by the generations before that, and now my generation is saying similar things to the next one. Of course, a decade or so later we always say, 'How could we have thought *that* was equality?' Are we serving the next generation well if we tell them that everything is equal and fair when it's not?[40]

In the second part of the book, 'Neurosexism', we take a closer look at claims about male and female brains. What do people *mean* when they say that there are inherent gender differences, or that the two sexes are hardwired to be better suited to different roles and occupations? As cognitive neuroscientist Giordana Grossi notes, these readily used phrases, 'along with the continual references to sex hormones, evoke images of stability and unchangeability: women and men behave differently because their brains are structured differently.'[41] Avid readers of popular science books and articles about gender may well have formed the impression that science has shown that the path to a male or a female brain is set in utero, and that these differently structured brains create essentially different minds. There *are* sex differences in the brain. There are also large (although generally decreasing) sex differences in who does what, and who achieves what. It would make sense if these facts were connected in some way, and perhaps they are. But when we follow the trail of contemporary science we discover a surprising number of gaps, assumptions, inconsistencies, poor methodologies, and leaps of faith – as well as more than one echo of the insalubrious past. As Brown University professor of biology and gender studies Anne Fausto-Sterling has pointed out, 'despite the many recent insights of brain research, this organ remains a vast unknown, a perfect medium on which to project, even unwittingly, assumptions about gender.'[42] The sheer complexity of the brain lends itself beautifully to overinterpretation and precipitous conclusions. After combing through the controversies, we'll ask whether modern neuroscientific explanations of gender inequality are doomed to join the same scrap heap as measures of skull volume, brain weight and neuron delicacy.

And it's important for scientists to remain aware of this possibility because from the seeds of scientific speculation grow the monstrous fictions of popular writers. Again and again, claims are made by so-called experts that are 'simply coating old-fashioned stereotypes with a veneer of scientific credibility', as Caryl Rivers and Rosalind Barnett warn in the *Boston Globe*.[43] Yet this 'popular neurosexism' easily finds its way into apparently scientific books and articles for the interested public, including parents and teachers.[44] Already, sexism disguised in neuroscientific finery is changing the way children are taught.

Neurosexism reflects and reinforces cultural beliefs about gender – and it may do so in a particularly powerful way. Dubious 'brain facts' about the sexes become part of the cultural lore. And, as I describe in 'Recycling Gender', the third part of the book, refreshed and invigorated by neurosexism, the gender cycle is ready to sweep up into it the next generation. Children, keen to understand and find their place in society's most salient social divide, are born into a half-changed world, to parents with half-changed minds.

I don't think that in my lifetime there will be a woman Prime Minister.

—Margaret Thatcher (1971), Prime Minister of
Great Britain from 1979 to 1990[45]

It's worth remembering just how much society can change in a relatively short period of time. Precedents are still being set. Could a society in which males and females hold equal places ever exist? Ironically, perhaps it is not biology that is the implacably resistant counterforce, but our culturally attuned minds.[46] No one knows whether males and females could ever enjoy perfect equality. But of this I am confident: So long as the counterpoints provided by the work of the many researchers presented in this book are given an audience, in fifty years' time people will look back on these early-twenty-first-century debates with bewildered amusement, and wonder how we ever could have thought that *that* was the closest we could get to equality.

PART I

'Half-Changed World',
Half-Changed Minds

> The more I was treated as a woman, the more woman I became.
> I adapted willy-nilly. If I was assumed to be incompetent at
> reversing cars, or opening bottles, oddly incompetent I found
> myself becoming. If a case was thought too heavy for me, inex-
> plicably I found it so myself.
>
> —Jan Morris, a male-to-female transsexual describing her post-
> transition experiences in her autobiography, *Conundrum* (1987)[1]

Suppose a researcher were to tap you on the shoulder and ask
you to write down what, according to cultural lore, males
and females are like. Would you stare at the researcher blankly
and exclaim, 'But what can you mean? Every person is a unique,
multifaceted, sometimes even contradictory individual, and with
such an astonishing range of personality traits within each sex,
and across contexts, social class, age, experience, educational level,
sexuality and ethnicity, it would be pointless and meaningless to
attempt to pigeonhole such rich complexity and variability into two
crude stereotypes'? No. You'd pick up your pencil and start writing.[2] Take a look at the two lists from such a survey, and you will
find yourself reading adjectives that would not look out of place in
an eighteenth-century treatise on the different duties of the two
sexes. One list would probably feature communal personality traits
such as *compassionate*, *loves children*, *dependent*, *interpersonally sensitive*, *nurturing*. These, you will note, are ideal qualifications for
someone who wishes to live to serve the needs of others. On the

other character inventory we would see agentic descriptions like *leader, aggressive, ambitious, analytical, competitive, dominant, independent* and *individualistic.* These are the perfect traits for bending the world to your command, and earning a wage for it.[3] I don't have to tell you which is the female list and which is the male one: you already know. (These lists, as sociologists Cecilia Ridgeway and Shelley Correll have pointed out, also most closely match stereotypes of 'white, middle-class, heterosexual men and women, if anyone.')[4]

Even if you, personally, don't subscribe to these stereotypes, there is a part of your mind that isn't so prissy. Social psychologists are finding that what we can consciously report about ourselves does not tell the whole story.[5] Stereotypes, as well as attitudes, goals, and identity also appear to exist at an implicit level, and operate 'without the encumbrances of awareness, intention, and control', as social psychologists Brian Nosek and Jeffrey Hansen have put it.[6] The implicit associations of the mind can be thought of as a tangled but highly organised network of connections. They connect representations of objects, people, concepts, feelings, your own self, goals, motives and behaviours with one another. The strength of each of these connections depends on your past experiences (and also, interestingly, the current context): how often those two objects, say, or that person and that feeling, or that object and a certain behaviour have gone together in the past.[7]

So what does the implicit mind automatically associate with women and men? The various tests that social psychologists use to assess implicit associations work from the assumption that if you present your participant with a particular stimulus, then this will rapidly, automatically and unintentionally activate strongly associated concepts, actions, goals and so on, more than weakly associated ones. These primed representations become more readily accessible to influence perception and guide behaviour.[8] In one of the most widely used tests, the computer-based Implicit Association Test or IAT (developed by social psychologists Anthony Greenwald, Mahzarin Banaji and Brian Nosek), participants must pair categories of words or pictures.[9] For example, first they might have to pair female names with communal words (like *connected* and *supportive*), and male names with agentic words (like *individualistic* and *competitive*). Participants usually find this easier than the opposite pairing (female names with agentic words, and male names with communal words). The small but significant difference in reaction time this creates is taken as a measure of the stronger automatic and unintended associations between women and communality, and men and agency.[10]

You probably have similar associations, regardless of whether you consciously endorse them. The reason for this is that the learning of these associations is also a process that takes place without the need for awareness, intention and control. The principle behind learning in associative memory is simple: as its name suggests, what is picked up are associations in the environment. Place a woman behind almost every vacuum cleaner being pushed around a carpet and, by Jove, associative memory will pick up the pattern. This certainly has its benefits – it's an effortless and efficient way to learn about the world around you – but it also has its drawbacks. Unlike explicitly held knowledge, where you can be reflective and picky about what you believe, associative memory seems to be fairly indiscriminate in what it takes on board. Most likely, it picks up and responds to cultural patterns in society, media and advertising, which may well be reinforcing implicit associations you don't consciously endorse. What this means is that if you are a liberal, politically correct sort of person, then chances are you won't very much like your implicit mind's attitudes. Between it and your conscious, reflective self there will be many points of disagreement. Researchers have shown that our implicit representations of social groups are often remarkably reactionary, even when our consciously reported beliefs are modern and progressive.[11] As for gender, the automatic associations of the categories male and female are not a few flimsy strands linked to penis and vagina. Measures of implicit associations reveal that men, more than women, are implicitly associated with science, maths, career, hierarchy and high authority. In contrast, women, more than men, are implicitly associated with the liberal arts, family and domesticity, egalitarianism and low authority.[12]

The results of a series of experiments by Nilanjana Dasgupta and Shaki Asgari at the University of Massachusetts give us an indication of how the media, and life itself, can give rise to these associations, quite independently of our consciously endorsed beliefs. These researchers looked at the effects of counterstereotypic information. In the first study, they gave one group of women a series of short biographies of famous women leaders to read (like Meg Whitman, then CEO of e-Bay, and Ruth Bader Ginsburg, US Supreme Court Justice). Afterwards, these women found it easier to pair female names with leadership words on the IAT, compared with controls who had not just read about women leaders. However, reading about these exceptional women had not an ounce of effect on the women's explicit beliefs about women's leadership qualities. Dasgupta and Asgari then went on to look at the effects of the real world on the implicit mind. They recruited women from two liberal arts colleges in the United States, one a women's college and the other coed. The researchers measured the women's implicit and conscious attitudes towards women and leadership during the first few months of freshman year and then again a year later. The type of college experience – coed or single sex – had no effect on the students' self-reported beliefs about women's capacity for leadership. However, it did have an effect on their implicit attitudes. At the beginning of freshman year, both groups of women were slow to pair female and leadership words on the IAT. But by sophomore year, the women at the single-sex college had lost this implicit disinclination to associate women with leadership, while coed students had become even slower at pairing such words. This divergence appeared to be due to students in women's colleges tending to have more exposure to female faculty, and coed students – particularly those who took maths and science classes – having less experience with women in leadership positions. The patterns of their environment, in other words, altered the gender stereotypes represented in the implicit mind.[13]

When gender is salient in the environment, or we categorise someone as male or female, gender stereotypes are automatically primed. For several years, social psychologists have been investigating how this activation of stereotypes affects our perception of others.[14] But more recently, social psychologists have also become interested in the possibility that sometimes we might also perceive our own selves through the lens of an activated stereotype. For, as it turns out, the self-concept is surprisingly malleable.

Perhaps, on presenting your psyche to a psychiatrist for analysis, you would fail to see a brightening of the eye, a gleam that anticipates an hour that is more pleasure than work. But even if your personality offers little to hold the interest of a shrink, there is nonetheless plenty in there to fascinate the social psychologist. This is because your self has multiple strings to its bow, it's a rich, complex web, it has a nuance for every occasion. As Walt Whitman neatly put it, 'I am large: I contain multitudes.'[15] But while a self that runs to the multitudes is certainly a fine thing to own, you can immediately see that it is not ideal to have the entire multitude in charge at the same time. What works better is if, at any one time, just a few self-concept items are plucked out from the giant Wardrobe of Self.

Some psychologists refer to whatever self is in current use – the particular self-concept chosen from the multitudes – as the active self.[16] As the name implies, this is no passive, sloblike entity that idles unchanging day after day, week after week. Rather, the active self is a dynamic chameleon, changing from moment to moment in response to its social environment. Of course, the mind can only make use of what is available – and for each of us certain portions of the self-concept come more easily to hand than do others. But in all of us, a rather large portion of the Wardrobe of Self is taken up with the stereotypical costumes of the many social identities each person has (New Yorker, father, Hispanic American, vet, squash player, man). Who you are at a particular moment – which part of your self-concept is active – turns out to be very sensitive to context. While sometimes your active self will be personal and idiosyncratic, at other times the context will bring one of your social identities hurtling towards the active self for use. With a particular social identity in place, it would not be surprising if self-perception became more stereotypic as a result. In line with this idea, priming gender seems to have exactly this effect.[17]

In one study, for example, a group of French high school students was asked to rate the truth of stereotypes about gender difference in talent in maths and the arts before rating their own abilities in these domains. So, for these students, gender stereotypes were very salient as they rated their own ability. Next, they were asked to report their scores in maths and the arts on a very important national standardised test taken about two years earlier. Unlike students in a control condition, those in the stereotype-salient group altered the memory of their own objective achievements to fit the well-known stereotype. The girls remembered doing better than they really had in the arts, while the boys inflated their marks in maths. They gave themselves, on average, almost an extra 3 percent on their real score while the girls subtracted the same amount from their actual maths score. This might not seem like a large effect, but it's not impossible to imagine two young people considering different occupational paths when, with gender in mind, a boy sees himself as an A student while an equally successful girl thinks she's only a B.[18]

If this method of priming gender doesn't seem very subtle, it's because it isn't. Of course that's not to say that it might not provide a useful proxy for the real world. Gender stereotypes are ubiquitous, sometimes even in settings where they shouldn't be. When the Scottish Qualifications Authority recently announced a drive to increase the dismally low numbers of senior school girls in subjects like physics, woodworking, and computing, some teachers freely expressed doubt that it was worth the effort. 'I think it is much better to realise that there are differences between boys and girls, and ways in which they learn', said a headmaster at a well-known Edinburgh private school. 'Overall, boys choose subjects to suit their learning style, which is more logic based'.[19] He was gracious enough to leave his audience to make the inference that girls' preferred learning style is an illogical one, rather than making the point explicitly. But importantly, gender identity can also be primed without the help of openly expressed stereotypes. Have you, for example, ever filled in a question on a form that looks something like this?

☐ Male
☐ Female

Even an innocently neutral question of this kind can prime gender. Researchers asked American university students to rate their mathematical and verbal abilities, but beforehand, some students were asked to note down their gender in a short demographics section, and others to mark their ethnicity.[20] The simple process of ticking a box had surprising effects. European American women, for example, felt more confident about their verbal skills when gender was salient (consistent with the prevailing belief that females have the edge when it comes to language skills) and rated their maths ability lower, compared with when they identified themselves as European American. In contrast, European American men rated their maths ability higher when they were thinking of themselves as men (rather than as European Americans), but their verbal ability better when their ethnicity had been made salient.

Even stimuli that are so subtle as to be imperceptible can bring about a change in self-perception. Psychologists Jennifer Steele and Nalini Ambady gave female students a vigilance task, in which they had to indicate with a key press, as quickly as possible, on which side of the computer screen a series of flashes appeared.[21] These flashes, were, in fact, subliminal primes: words replaced so quickly by a string of Xs that the word itself couldn't be identified. For one group, the words primed 'female' (*aunt, doll, earring, flower, girl* and so on). The other group saw words like *uncle, hammer, suit, cigar* and *boy*. Then, the volunteers were asked to rate how much pleasure they found in both feminine activities (like writing an essay or taking a literature exam) and masculine tasks (like solving an equation, taking a calculus exam or computing compound interest). The male-primed group of women rated both types of activity as equally enjoyable. But the female-primed group reported a preference for arts-related activities over maths-based ones. The prime 'changed women's lens of self-perception', the authors suggest.[22]

We are not just influenced by the imperceptible, but also the intangible. The Australian writer Helen Garner noted that one can either 'think of people as discrete bubbles floating past each other and sometimes colliding, or . . . see them overlap, seep into each other's lives, penetrate the fabric of each other'.[23] Research supports the latter view. The boundary of the self-concept is permeable to other people's conceptions of you (or, somewhat more accurately, your perception of their perception of you). As William James put it, 'a man has as many social selves as there are individuals who recognise him and carry an image of him in their mind.'[24] By way of scientific support for James's idea, Princeton University psychologist Stacey Sinclair and her colleagues have shown in a string of experiments that people socially 'tune' their self-evaluations to blend with the opinion of the self held by others. With a particular person in mind, or in anticipation of interacting with them, self-conception adjusts to create a shared reality. This means that when their perception of you is stereotypical, your own mind follows suit. For example, Sinclair manipulated one group of women into thinking that they were about to spend some time with a charmingly sexist man. (Not a woman-hater, but the kind of man who thinks that women deserve to be cherished and protected by men, while being rather less enthusiastic about them being too confident and assertive.) Obligingly, the women socially tuned their view of themselves to better match these traditional opinions. They regarded themselves as more stereotypically feminine, compared with another group of women who were expecting instead to interact with a man with a more modern view of their sex.[25] Interestingly, this social tuning only seems to happen when there is some sort of motivation for a good relationship. This suggests that close or powerful others in your life may be especially likely to act as a mirror in which you perceive your own qualities.

These shifts in the self-concept do not just bring about changes in the eye of the self-beholder. They can also change behaviour. In her report of kindergarten children, sociologist Bronwyn Davies describes how one little girl, Catherine, reacts when the doll she is playing with is snatched away by a boy. After one failed attempt to retrieve the doll, Catherine strides to the dress-up cupboard and pulls out a man's waistcoat. She puts it on, and 'marches out. This time she returns victorious with the dolly under her arm. She immediately takes off the waistcoat and drops it on the floor.'[26] When adults pull a new active self out of the wardrobe, the change of costume is merely metaphorical. But might it nonetheless, as it did for Catherine, help us better fulfil a particular role or goal? Research suggests that it can.

In a recent series of experiments, Adam Galinsky at North-western University and his colleagues showed participants a photograph of someone: a cheerleader, a professor, an elderly man, or an African American man. In each case, some of the volunteers were asked to pretend to actually *be* the person in the photograph and to write about a typical day as that person. Control participants were told to write about a typical day in the person's life from a more dispassionate, third-person (he/she . . .) point of view. (This meant the researchers could see the effects of perspective-taking over and above any effects of priming a stereotype.) The researchers discovered that perspective-taking gave rise to 'self-other merging'. Asked to rate their own traits after the exercise, those who had imagined themselves as a cheerleader rated themselves as more attractive, gorgeous and sexy, compared with controls. Those who imagined themselves as professors felt smarter, those who walked in the shoes of the elderly felt weaker and more dependent, and those who had temporarily lived life as an African American man rated themselves as more aggressive and athletic. Self-perception absorbed the stereotypical qualities of another social group.[27]

The researchers then went on to show that these changes in the self-concept had an effect on behaviour. Galinsky and his colleagues found that pretending to be a professor improved analytic skills compared with controls, while a self-merging with cheerleader traits impaired them. Those who had imagined themselves as an African American man behaved more competitively in a game than those who had briefly imagined themselves to be elderly. The simple, brief experience of imagining oneself as another transformed both self-perception and, through this transformation, behaviour. The maxim 'fake it till you make it' gains empirical support.

No less remarkable effects on behaviour were seen by Stacey Sinclair and her colleagues. You'll recall that women who thought they were about to meet a man with traditional views of women perceived themselves as more feminine than women who expected to meet a man with more modern opinions. In one experiment, Sinclair arranged for her participants to actually interact with this man. (Of course, he was really a stooge, but didn't know what each woman thought he thought about women.) Women who thought he was a benevolent sexist didn't just think themselves more feminine, they also behaved in a more stereotypically feminine way.[28] (As a psychologist who has worked for several years in philosophy departments, perhaps this is a good moment to suggest to any colleagues who have found tearoom conversations with me intellectually unsatisfying that they have only their low opinion of psychologists to blame.)

It's not hard to see just how useful and adaptable a dynamic sense of self can be.[29] As the pivot through which the social context – which includes the minds of others – alters self-perception, a changing social self can help to ensure that we are wearing the right psychological hat for every situation. As we've begun to see, this change in the self-concept can then have effects for behaviour, a phenomenon we'll look at more closely in the chapters that follow. With the right social identity for the occasion or the companion, this malleability and sensitivity to the social world helps us to fit ourselves into, as well as better perform, our current social role. No doubt the female self and the male self can be as useful as any other social identity in the right circumstances. But flexible, context-sensitive and useful is not the same as 'hardwired'. And, when we take a closer look at the gender gap in empathising, we find that what is being chalked up to hardwiring on closer inspection starts to look more like the sensitive tuning of the self to the expectations lurking in the social context.

WHY YOU SHOULD COVER YOUR HEAD WITH A PAPER BAG IF YOU HAVE A SECRET YOU DON'T WANT YOUR WIFE TO FIND OUT

2

> One morning at breakfast, my patient Jane looked up to see that her husband, Evan, was smiling. He held the newspaper, but his gaze was lifted and his eyes darted back and forth, though he wasn't looking at her. She had seen this behavior many times before in her lawyer husband and asked, 'What are you thinking about? Who are you beating in court right now?' Evan responded, 'I'm not thinking about anything.' But in fact he was unconsciously rehearsing an exchange with counsel he might be having later that day – he had a great argument and was looking forward to mopping up the courtroom with his opponent. Jane knew it before he did.
>
> —Louann Brizendine, *The Female Brain* (2007)[1]

Goodness, but Brizendine sets the bar high for women. I am trying in vain to recall an occasion during our many years together when, glancing up to see my husband's fingers twitching over the cereal bowl, I startled him by presciently asking, 'What are you thinking about? What invoice are you paying right now?' To be brutally honest, at breakfast I prefer to reserve the majority of my neurons for the thinking of my own thoughts, not those of others. But while Brizendine's claims are somewhat extravagant – is it *really* true that women have more privileged access to men's thoughts than they do themselves, or that 'a man can't seem to spot an emotion unless someone cries or threatens bodily harm'?[2] – we're all familiar with the concept of womanly intuition and womanly tenderness.

It's important, by the way, not to jumble together these two distinct 'feminine' skills. When a man looks for a soul mate to refresh his overlaboured faculties and unbend his learned brow, if he is wise he will check for two different qualities in his potential candidates. First, he needs someone who is quick to discern – from, for example, its furrowed appearance – that his brow is indeed in need of straightening. This is cognitive empathy, the ability to intuit what another person is thinking or feeling. But in addition, she needs to be the kind of person who will use her powers of interpersonal perception for good, not evil. Affective empathy is what we commonly think of as sympathy – feeling and caring about the other person's distress. Put the two together and you have an angel in human form. As Baron-Cohen describes it in *The Essential Difference*, 'imagine you not only see Jane's pain, but you also automatically feel concern, wince, and feel a desire to run across and help alleviate her pain.'[3]

As we already know, according to Baron-Cohen it is women on average who are 'predominantly hard-wired' to see, feel, wince, run and alleviate. His Empathy Quotient (or EQ) questionnaire asks people to report their skill and inclination for both cognitive and affective empathy with statements like *I can easily tell if someone else wants to enter a conversation* and *I really enjoy caring for other people*. (The person filling in the questionnaire agrees or disagrees, slightly or strongly, with each statement.) To diagnose what he calls brain sex, Baron-Cohen uses the EQ together with its brother the Systemising Quotient (SQ), which poses questions like *If there was a problem with the electrical wiring in my home, I'd be able to fix it myself* and *When I read the newspaper, I am drawn to tables of information, such as football league scores or stock market indices*.[4] People who score higher on the EQ than the SQ have an E-type or female brain, and the opposite result indicates an S-type or male brain.

The large minority who score approximately equally on the two tests are deemed to have a balanced brain. Baron-Cohen reports that just under 50 percent of women, but only 17 percent of men, have a female brain.[5]

As journalist Amanda Schaffer pointed out in *Slate* there is something curious about equating empathising with the female brain when, albeit by a whisker, the majority of women do *not* claim to have a predominantly empathising focus. She reports that when she asked Baron-Cohen about this, he 'admitted that he's thought twice about his male brain/female brain terminology, but he didn't disavow it.'[6] And, while we're on the subject of terminology, calling a test the 'Empathy Quotient' does not, on its own, make it a test of empathising. Asking people to report on their own social sensitivity is a bit like testing mathematical ability with questions like *I can easily solve differential equations*, or assessing motor skills by asking people to agree or disagree with statements like *I can pick up new sports very quickly*. There's something doubtfully subjective about the approach.

As it turns out, doubt is well-justified, for both affective and cognitive empathy. In an important review of gender differences in affective empathy, psychologists Nancy Eisenberg and Randy Lennon found that the female empathic advantage becomes vanishingly smaller as it becomes less and less obvious that it is something to do with empathy that is being assessed.[7] (So, gender differences were greatest on tests in which it was very clear what was being measured, that is, on self-report scales. Smaller differences were seen when the purpose of the testing was less obvious. And no gender difference was found for studies using unobtrusive physiological or facial/gestural measures as an index of empathy.) In other words, women and men may differ not so much in actual empathy but in 'how empathetic they would like to appear to others (and, perhaps, to themselves)', as Eisenberg put it to Schaffer.[8]

As for cognitive empathy there is, it appears, no shortage of people in the world who can unwittingly offend, misunderstand and steamroller over the delicate signals of others, all while maintaining the self-perception that they are unsurpassedly sensitive to subtle social cues. When psychologists Mark Davis and Linda Kraus analysed all the then-relevant literature in search of an answer to the question, what makes for a good empathiser? their conclusion was surprising. They found that people's ratings of their own social sensitivity, empathy, femininity and thoughtfulness are virtually useless when it comes to predicting actual interpersonal accuracy. As the authors conclude, 'the evidence thus far leaves little doubt that traditional self-report measures of social sensitivity have minimal value in allowing us to identify good or poor judges.'[9] A more recent study 'found only weak or non-significant correlations between self-estimates of performance and actual performance', while another, with a sample of more than 500 participants, supported the 'still surprising conclusion that people, in general, are not very reliable judges of their own mind-reading abilities.'[10]

A few studies have found links between self-perception of empathising skill and actual ability, I should note. Recently, a large Austrian study of more than 400 people found that EQ score correlated modestly with something called the Reading the Mind from the Eyes test.[11] (In this multiple-choice test, the participant is shown just the eye region of a series of faces and asked to guess each person's mental state.) But this relationship is the exception rather than the rule. (And in this case, there might be an unexpected reason for the link.)[12] As an expert on the subject of empathy, University of Texas–Arlington professor William Ickes, suggested in his book *Everyday Mind Reading*, 'most perceivers may lack the kind of metaknowledge they would need to make valid self-assessments of their own empathic ability',[13] which is a politely academic way of saying that if you want to predict people's empathic ability you might as well save everyone's time and get monkeys to fill out the self-report questionnaires. And so to find, as Baron-Cohen does, that women score relatively higher on the EQ is not terribly compelling evidence that they *are*, in fact, more empathic. Nor is it hard to come up with a plausible hypothesis as to why they might give themselves undeservedly higher scores. As we saw in the previous chapter, when the concept of gender is primed, people tend to perceive themselves in more stereotypical ways. The statements in the EQ could conceivably prime gender on their own. As philosopher Neil Levy has pointed out, the statements in the EQ and SQ are 'often testing for the gender of the subject, by asking whether the subject is interested in activities which tend to be disproportionately associated with males or with females (cars, electrical wiring, computers and other machines, sports and stock markets, on the one hand, and friendships and relationships, on the other).'[14] And in any case, the questionnaire asks participants to note their sex before filling in the questionnaire, which we know can prime a gender identity. So are women *actually* better at guessing other people's thoughts and feelings?

The idea of womanly intuition isn't without empirical support. In the Austrian study, women scored higher than men on the Reading the Mind from the Eyes test. However, the difference was small: women, on average, correctly guessed 23 of the 36 items; men, 22.[15] Women also score reliably, if modestly, higher than men on a test called the Profile of Nonverbal Sensitivity (PONS). In this test the participant watches a woman acting out a series of very short, and very stripped down, scenes. Each scene is just two seconds long, and the viewer sees only a few channels of information: such as only the body and hands, or just the face. From this minimal information, the viewer has to choose one of two possible descriptions of the scene.[16] Yet despite women's slight advantage on the PONS overall, the detailed picture is a little more nuanced. At a dinner party, when you listen to someone explain the system they have discerned in the latest football league scores, you are easily able to convey your fascination by way of a polite smile. But the so-called leaky channels of communication – for example, your body language and fleeting microexpressions – are less readily controlled. On the PONS, women are particularly adept at decoding the most controlled forms of communication, like facial expression, but, the leakier the channel, the smaller their advantage.

This is odd. Isn't women's intuition supposed to specialise in the hidden stuff other people can't see? Brizendine, for example, describes women's intuition as an ability to 'feel a teenage child's distress, a husband's flickering thoughts about his career, a friend's happiness in achieving a goal, or a spouse's infidelity at a gut level.'[17] But it now seems that womanly intuition is the authority in posed feelings rather than the perhaps more interesting true emotions that leak out in other ways. One explanation put forward for this is that women are socialised to be polite decoders who would as soon peer through the keyhole of an occupied restroom stall as scrutinise someone's unintended emotional leaks.[18]

What's more, tests like the Reading the Mind in the Eyes task and the PONS are not exactly what you would call realistic simulations of everyday mind reading. Trying to penetrate the expression of the *Mona Lisa*, or talking to a time-pressed Muslim woman in full burka might come close to what they assess – but arguably, social interactions more typically involve a stream of rich and changing information from other people (who do not offer multiple-choice options as to what they might be feeling). In the 1990s, William Ickes and his colleagues developed a new empathy test, one Ickes probably rightly claims is 'the most stringent test' of a person's ability to infer the thoughts and feelings of others.[19] In this empathic accuracy test, two people wait together for an experiment to begin. The experimenter has departed to find a replacement for the projector bulb that has just blown – and in fact, the experiment has already begun. As they sit there and wait, they are unobtrusively filmed and recorded for six minutes. On her return, the experimenter explains the true purpose of the experiment. If both parties are happy to continue, they then view the film clip of their interaction individually, and as they go through the tape they pause it every time they recall having had a specific thought or feeling, and jot down what it was, and whether it was positive, negative, or neutral. Then, in the last part of the experiment, each person watches the tape again, but this time it's stopped every time the *partner* reported a feeling or thought. The task is to infer what this was. This can then be compared to what the partner actually reported feeling or thinking at that very moment.

You will probably agree that, of all the tests mentioned so far, this seems to most closely approximate real-world empathising. There are no actors posing expressions, no narrow strips of eyes, no disembodied voices and hands, no carefully choreographed and scripted scenes. Instead, people are interacting in a natural and unscripted way that generates a stream of successive mental states to be inferred from a rich variety of clues. You might expect men to struggle with such a demanding test, but they do not. As Ickes reports in *Everyday Mind Reading*, much to everyone's surprise, in the first seven studies to use this measure no gender differences were found:

> Where was the empathic advantage that we commonly refer to as 'women's intuition?' It wasn't evident in the interactions of opposite-sex strangers, or in the interactions of heterosexual dating partners, or even in the interactions of recently-married or longer-married dating partners. It wasn't evident in comparisons of female-female dyads with male-male dyads or of all-female groups with all-male groups. It wasn't evident in Texas, in North Carolina, or even in New Zealand. Was it nothing more than a cultural myth? A fictitious bit of folklore that was ripe for scientific debunking?

But then, something 'baffling' happened.[20] The next three studies, all of which took place four or more years after the first empathic accuracy study, *did* find gender differences. The researchers quickly spotted that there had been a slight change in the form that the viewers used while going through the tape of the interaction. In the new form, for each thought and feeling that they guessed, they had to say how accurate they thought they were. When this version of the form was used womanly intuition existed; when the old form was used, it didn't.[21] Why might this be? Ickes suggested that this small change reminds women that they *should* be empathic, and therefore increases their motivation on the task. He concludes from his lab's research that '[a]lthough women, on average, do not appear to have more empathic *ability* than men, there is compelling evidence that women will display greater accuracy than men when their empathic *motivation* is engaged by situational cues that remind them that they, as women, are expected to excel at empathy-related tasks.'[22]

If so, then if the experimental situation can instead be designed to motivate *men*, then their empathic performance should also improve. This is exactly what researchers are beginning to find. Kristi Klein and Sara Hodges used an empathic accuracy test in which participants watched a video of a woman talking about her failure to get a high enough score on an exam to get into the graduate school she wanted to attend.[23] When the feminine nature of the empathic accuracy test was highlighted by asking participants for sympathy ratings before the empathic accuracy test, women scored significantly better than men. But a second group of women and men went through exactly the same procedure but with one vital difference: they were offered money for doing well. Specifically, they earned $2 for every correct answer. This financial incentive levelled the performance of women and men, showing that when it literally 'pays to understand' male insensitivity is curiously easily overcome.

You can also improve men's performance by inviting them to see a greater social value in empathising ability. Cardiff University psychologists presented undergraduate men with a passage titled 'What Women Want'.[24] The text, complete with bogus references, then went on to explain that contrary to popular opinion 'non-traditional men who are more in touch with their feminine side' are regarded as more sexually desirable and interesting by women, not to mention more likely to leave bars and clubs in the company of one. Men who read this passage performed better on the empathic accuracy task than did control men (to whom the test was presented in a nothing-to-do-with-gender fashion) or men who had been told that the experiment was investigating their alleged intuitive inferiority.

Clearly, one's performance on cognitive empathy tasks involves a combination of motivation and ability. If social expectations can create a motivation gap, could they also be responsible for an ability gap? Women on average score better than men on another social sensitivity test called the Interpersonal Perception Task (IPT). Here, participants watch and hear people acting out unscripted interactions. From the actors' verbal and nonverbal behaviour, the viewers have to try to work out the nuances of their relationships. For example, from watching a scene between two men and a child, the participant has to work out which man is the child's father. Recently, psychologists Anne Koenig and Alice Eagly used the IPT to explore the idea that the gender stereotype of women's superior social skills might furnish women with an unfair advantage.[25] To one group, the test was accurately described as a measure of social sensitivity, or 'how well people accurately understand the communication of others and the ability to use subtle nonverbal cues in everyday conversations.' Before the participant took the test, the experimenter casually mentioned that 'We've been using this test for a couple of quarters now. It's 15 questions long and, not surprisingly, men do worse than women.' In this group, the men did indeed do slightly worse than the women. But to a second group of participants, the test was described in a more gender-neutral way. It was presented as a measure of complex information processing, or 'how well people process different kinds of information accurately.' In this group, the men performed just as well as the women.

The take-home message of these studies is that we can't separate people's empathising ability and motivation from the social situation. The salience of cultural expectations about gender and empathising interacts with a mind that knows to which gender it belongs. So what would happen if we could temporarily trick a female mind into thinking it was male? As we saw in the previous chapter, when people take the first-person, 'I' perspective of someone else, the stereotypical traits of the other permeate and seep into the perspective-takers' own self-concept. This merging of identities can cross gender boundaries. A few years ago, psychologists David Marx and Diederik Stapel asked a group of Dutch undergraduates to write about a day in the life of a student named Paul. Half of the students wrote in the first-person ('I') while the other half used the third-person perspective ('he'). Afterwards, they were asked to rate themselves on technical-analytic skills and emotional sensitivity skills. For the female undergraduates, thinking of themselves as Paul in the 'I' perspective altered their self-conceptions. Women who attempted to be Paul living his life actually incorporated his stereotypical male characteristics into their own self-conceptions. They rated themselves as higher on analytic abilities and lower on emotional sensitivity, compared with women who had written a third-person story. In other words, there was 'a merging between the self and [Paul], such that female participants became more "malelike" as a result.'[26] Indeed, they became so malelike that their self-ratings on these stereotypical traits were statistically indistinguishable from the men's. For men, there was no such effect of being Paul on their self-concept, presumably because they already *were* a male student.

The participants were also given a battery of emotion sensitivity tests. These problems included ones like recognising facial expressions of emotion, choosing which two more basic emotions make up more complex ones (like optimism), and working out, for example, what emotional state you reach as you become more and more guilty and lose your feeling of self-worth. (Is it *depression*, *fear*, *shame* or *compassion*?) Women who had not put themselves in male shoes performed a lot better than the men on this task, getting on average 72 percent of the emotion-sensitivity questions correct, while men's scores hovered around the 40 percent mark. But women who had just spent a few moments merely imagining themselves to be a man performed every bit as poorly as the real men.

No doubt an intricate interplay between minds and social expectations affects our capacity for affective empathy, too. Research into group-based emotions investigates the idea that when 'people are thinking of themselves in terms of a particular group membership – whenever a social rather than personal identity is salient – people's emotional experiences and reports will be shaped and determined by that group membership.'[27] In a recent study, researchers found that subtly priming a social identity led people to experience group-based emotions that were different from those they experienced when thinking of themselves as an individual. Is it possible that women become more tenderhearted when thinking of themselves as women or mothers rather than as individuals or, say, saleswomen?

We don't know, but University of Exeter psychologist Michelle Ryan and her colleagues have found that the social identity you are wearing can certainly change the sway of compassionate feeling in resolving moral dilemmas.[28] In the 1980s, Carol Gilligan famously suggested that women and men reason about moral situations in a different way. She suggested that the 'ethic of justice' – which privileges abstract principles of justice such as equality, reciprocity and universal rules – is used more by men. In contrast, the 'ethic of care' – which takes greater account of the feelings and relationships of those concerned – is used mostly by women. Researchers since have argued that what kind of ethic is used depends a great deal on who the moral dilemma involves: men and women alike are happy to apply abstract universal laws and principles to strangers, but tend to turn to the ethic of care for answers when considering the plight of friends or other intimates.[29] And any remaining gender difference in moral reasoning appears not to be hardwired, because it can be eliminated with a change of social identity. Ryan and colleagues presented students from the Australian National University (ANU) with a moral dilemma: A student from the local TAFE (a nonuniversity institute of tertiary education) urgently needs a book for an assignment due the next day. Without the book, the student will fail. The book is not available at the desperate student's own library. The ANU students are asked whether they would borrow the book from their own library, on behalf of the TAFE student.

Before being presented with this realistic dilemma, the researchers manipulated which social self was in charge by asking participants to brainstorm ideas for a debate. Then they were given the dilemma to read, and asked to explain the important factors involved and what they would do in that situation. One group was primed with gender stereotypes (they were asked to come up with debating ideas for the claim that men are still real men or that women are not the weaker sex). Within this group, there was clear evidence of gender difference in moral-reasoning style. Women were twice as likely to offer care-based considerations, such as the alleviation of another's suffering. This might lead us to think that men are less empathic in their approach to moral dilemmas − except that in two other groups, both primed with a student identity, gender made no difference. The second group of students was primed to think of themselves as tertiary students. With this identity in place, the TAFE student was one of them. The last group was primed with only their more exclusive identity as ANU students. (The Australian National University is arguably the highest ranking in the country.) Regardless of sex, tertiary-student-primed students offered more care-based considerations and fewer justice arguments than the ANU-primed students, who had been primed to feel socially distant to the harried TAFE student.

In other words, when we are not thinking of ourselves as 'male' or 'female', our judgements are the same, and women and men alike are sensitive to the influence of social distance that, rightly or wrongly, pushes moral judgements in one direction or another along the care-justice continuum. But moral reasoning is also sensitive to another social factor − the salience of gender. Thus, the authors argue that 'it is the salience of gender and gender-related norms, rather than gender per se, that lead to differences between women and men.' Of course, as they also point out, 'the social reality is that gender, for most, is a ubiquitous category and is arguably the most salient'.[30]

Let's rejoin Jane and Evan at the breakfast table and take stock for a moment.

In the eighteenth century Thomas Gisborne observed with pleasure how nature had conveniently endowed the female mind with those very qualities she most needed to discharge her social duties. Nowadays, the argument plays the other way: women choose the social roles that best fit their female mind. But perhaps Gisborne was nearly right, after all. The mind, triggered by social cues, uses its female identity to endow *itself* with the greater sensitivity, sympathy, and compassion ascribed to it by cultural belief. Then, just as remarkably, these enhancements are gone. It's as good as magic. But as we'll see in the next chapter, social psychology is full of these now-you-see-it, now-you-don't tricks.

Pick a gender difference, any difference. Now watch very closely as – *poof!* – it's gone.

Social psychologists are becoming rather brilliant at setting up these gender difference sleights of hand. The examples are piling up in all sorts of domains – from social sensitivity to chess to negotiation – but the pièce de resistance is the visuospatial skill of mental rotation performance.

In the classic and most widely used test of this ability, the test taker is shown an unfamiliar three-dimensional shape made up of little cubes – the target – and four other similar shapes. Two of these are the same as the original but have been rotated in three-dimensional space, and two are mirror images. The task is to work out which two are the same as the target. Mental rotation performance is the largest and most reliable gender difference in cognition. In a typical sample, about 75 percent of people who score above average are male.[1] Gender differences in mental rotation ability have even recently been seen in babies three to four and five months of age.[2] While it's easy to see that a high score on the mental rotation test would be a distinct advantage when it comes to playing Tetris, some also claim (although they're often strongly disputed) that male superiority in this domain plays a significant role in explaining males' better representation in science, engineering and maths.[3]

People's mental rotation ability is malleable; it can be greatly enhanced by training.[4] But there are far quicker, easier ways to modulate mental rotation ability. By now, you already know what these methods involve: manipulating the social context in such a way that it changes the mind that is performing the task. For example, you can feminise the task. When, in one study, participants were told that performance on mental rotation is probably linked with success on such tasks as 'in-flight and carrier-based aviation engineering . . . nuclear propulsion engineering, undersea approach and evasion, [and] navigation', the men came out well ahead. Yet when the same test was described as predicting facility for 'clothing and dress design, interior decoration and interior design . . . decorative creative needlepoint, creative sewing and knitting, crocheting [and] flower arrangement', this emasculating list of activities had a draining effect on male performance.[5]

Alternatively, instead of changing the gender of the task, you can keep the task the same but push gender into the mental background. Matthew McGlone and Joshua Aronson, for example, measured mental rotation ability in students at a selective liberal arts college in the northeastern United States. One group was primed with gender, while another group was primed with their exclusive private-college identity. Women who had been induced to think of themselves as a student at a selective liberal arts college enjoyed a performance boost, scoring significantly higher than gender-primed women.[6] Likewise, Markus Hausmann and colleagues found that although gender-stereotype-primed men outperformed gender-stereotype-primed women, men and women primed with an irrelevant (geographical region-based) stereotype performed similarly on the mental rotation task.[7]

Another outrageous, but successful, approach was recently devised by Italian researcher Angelica Moè.[8] She described the mental rotation test to her Italian high school student participants as a test of spatial abilities and told one group that 'men perform better than women in this test, probably for genetic reasons.' The control group was given no information about gender. But a third group was presented with a downright lie. That group was told that 'women perform better than men in this test, probably for genetic reasons.' So what effect did this have? In both the men-are-better and the control group, men outperformed women with the usual size of gender difference. But women in the women-are-better group, the recipients of the little white lie, performed just as well as the men.

How can such easy manoeuvres – changing the way a task is described, bringing a particular social identity to the fore, or telling a simple fib – have such an erosive effect on the most robust gender difference in cognition in the literature? We saw in the previous chapter that the social demands of a situation can change how motivated men and women are to perform well. And psychologists are beginning to uncover other ways in which the social context can change, for better or for worse, the mind's power and effectiveness. There turn out to be a striking number of ways that being in the 'wrong' social group creates a trickier psychological path to navigate. With regards to gender, researchers have had quite a lot of success unravelling how the social context interacts with ability in traditionally masculine domains, especially mathematics. As we'll see in this chapter, a female doing traditionally male work faces the same problem as the dancer Ginger Rogers, who, as it was once famously noted, 'did everything Fred Astaire did, except backwards and in high heels'.

In her history of American women in medicine, *Sympathy and Science*, Regina Markell Morantz-Sanchez relates the memorable operating room experience of an early-twentieth-century medical student, Mary Ritter:

> As the gruesome operation proceeded I gritted my teeth, clenched my hands, and held on. Next to me stood a senior woman student. I watched her turn a greenish white and sway a little. Contrary to the ethics of an operating room, where silence is the rule, I hissed in her ear, 'Don't you dare faint.' . . . The two women students did not faint and thus

disgrace the sex. That three men did faint was merely due to a passing circulatory disturbance of no significance; but had the two women medical students fainted, it would have been incontrovertible evidence of the unfitness of the entire sex for the medical profession.[9]

Ritter, as a female interloper in the mostly male domain of medicine, was acutely aware of what today is referred to as stereotype threat (or, sometimes, social identity threat), the 'real-time threat of being judged and treated poorly in settings where a negative stereotype about one's group applies'.[10] A now-substantial literature shows that, as in the mental rotation examples described earlier, changing the threat level of the context can have a tangible effect on ability.[11] One very striking and real-world demonstration of this was provided by City University of New York psychologist Catherine Good and colleagues, who used as their participants more than 100 university students enrolled in a fast and difficult calculus class that was a pipeline to the hard sciences.[12] The students were given a calculus test made up of questions from the Graduate Record Examination (GRE) Maths test and, to motivate them, were told that they would get extra credit based on their performance. (In reality, everyone received the same credit.) The test packet handed out to each student included some information about the test. Students in the stereotype threat condition were told that the test was designed to measure their maths ability, to try to better understand what makes some people better at maths than others. This kind of statement can on its own create stereotype threat for women, who are well aware of their own stereotyped inferiority in mathematics.[13] But added to this, in the nonthreat condition, was the information that despite testing on thousands of students no gender difference had ever been found. So what was the effect of this extra information?

The men and women in the two groups had, on average, all received much the same course grades. You'd expect then, given their apparently equivalent ability, that males and females in the threat and nonthreat condition would perform at about the same level on the test. Instead, the researchers found that females performed better in the nonthreat condition, and this was particularly striking among Anglo-American participants, who generally show the greatest sex difference in maths performance. Among these participants, men and women in the threat condition, as well as men in the nonthreat condition, all scored about 19 percent on this very difficult test. But women in the nonthreat group scored an average of 30 percent correct, thus outperforming every other group – including both groups of men. In other words, the standard presentation of a test seemed to suppress women's ability, but when the same test was presented to women as equally hard for men and women, it 'unleashed their mathematics potential.'[14]

It's disconcerting to think that those who belong to negatively stereotyped groups might be pervasively hampered by stereotype threat effects in their academic lives. Recently, Stanford University's Gregory Walton and his colleague Steven Spencer analysed data from dozens of stereotype threat experiments to test the idea that negatively stereotyped students' real-world academic performance is 'like the time of a track star running into a stiff headwind: It underestimates her time without the headwind.' They confirmed that negatively stereotyped participants (that is, females doing maths and non-Asian minority students), matched on real-world academic tests like the SAT, performed worse than nonstereotyped groups under stereotype threat. But importantly, when stereotype threat was removed, the stereotyped groups actually *outperformed* nonstereotyped peers who, from real-world tests, one would think had the same ability.[15]

Psychologists have been very creative in working out how stereotype threat can have such a dampening effect on performance. Occasionally, psychologists make up their own negative stereotypes. But mostly, they are content to exploit preexisting cultural beliefs about group differences, like women's inferior mathematical ability. This can be done in disquietingly naturalistic ways. Stereotype threat effects have been seen in women who: record their sex at the beginning of a quantitative test (which is standard practice for many tests); are in the minority as they take the test; have just watched women acting in air-headed ways in commercials, or have instructors or peers who hold – consciously or otherwise – sexist attitudes.[16] Indeed, subtle triggers for stereotype threat seem to be more harmful than blatant cues,[17] which suggests the intriguing possibility that stereotype threat may be more of an issue for women now than it was decades ago, when people were more loose-lipped when it came to denigrating female ability.

So what happens to the female mind under threat? Somewhat inconveniently, when faced with the prospect of a maths test that will probe one's mathematical strengths and weaknesses, the female mind brings out its gender identity.[18] The stereotype that females are poor at maths is now officially self-relevant, and this seems to be important. This might be why the private-college-primed women in Matthew McGlone's mental rotation study performed better than their gender-primed counterparts: the former were construing themselves as members of an intellectually elite establishment, rather than women. Research suggests that the deadly combination of 'knowing-and-being' (*women are bad at maths* and *I am a woman*) can lower performance expectations, as well as trigger performance anxiety and other negative emotions.[19] For example, Mara Cadinu and her colleagues at the University of Padova gave women a maths test similar to the Graduate Record Examination. Beforehand, some women were told that 'recent research has shown that there are clear differences in the scores obtained by men and women in logical-mathematical tasks', while the other participants were told that there were no such differences.[20] Before each of the problems in the test, the women were given a blank page on which they were asked to write down anything that popped into their heads. Women in the stereotype threat condition listed more than twice as many negative thoughts about the maths test (like, 'These exercises are too difficult for me'). As this negativity built up, it increasingly interfered with performance. Although in the first half of the test both groups scored on average around 70 percent, by the latter half of the exercise the control group's performance had slightly improved (to 81 percent) whereas the threat group's performance had plummeted to 56 percent.

Recently, Christine Logel and her colleagues found evidence that the mind struggles to suppress the negative stereotype-based thoughts activated by the situation.[21] She found that women interrupted just as they began a challenging maths test were actually *slower* than men to respond to words like *illogical*, *intuitive* and *irrational*. This was a sign that worried thoughts about being femininely illogical, intuitive, and irrational were being suppressed. A quirk of suppressed thoughts is that, afterwards, they become hyperaccessible. Sure enough, women tested immediately after the test was over were especially fast at responding to the stereotypical words. (By contrast, no such turmoil appeared to be taking place in the minds of the men.) Although you might think that suppressing negative stereotypical thoughts would help women, it doesn't. Logel found that the more women suppressed irrational-woman concepts, the worse they performed. The reason for this seems to be because suppressing unwanted thoughts and anxieties uses up mental resources that could be put to better use elsewhere. To perform well in a demanding mental task you have to remain focused. This involves keeping accessible the information you need for your computations, as well as keeping out of consciousness anything that is irrelevant or distracting. This mental housekeeping is the duty of what is known as working memory or executive control. Most people facing a difficult and important intellectual challenge are likely to have a few intrusive self-doubts and anxieties. But as we've seen, people performing under stereotype threat have more. This places an extra load on working memory – to the detriment of the cognitive feat you are trying to achieve.[22] Women (and others) under stereotype threat may also try to control the anxious emotions that accompany their negative thoughts, which, unfortunately, can further deplete working memory resources.[23]

As you will begin to appreciate, a mind that is struggling with negative stereotypes and anxious thoughts is not in a psychologically optimal state for doing taxing intellectual tasks. And it's important to bear in mind that these jittery, self-defeating mechanisms are not characteristic of the *female* mind – they're characteristic of the mind *under threat*. Similar effects have been seen in other social groups put under stereotype threat (including white men).[24] And when researchers make the test-taking situation less threatening to women – that is, attempt to create for them the kind of situation in which men usually take maths tests – they don't see these negative effects on working memory and performance.[25]

In addition to clogging up working memory, stereotype threat can also handicap the mind with a failure-prevention mindset. The mind turns from a focus on seeking success (being bold and creative) to a focus on avoiding failure, which involves being cautious, careful, and conservative (referred to as *promotion focus* and *prevention focus*, respectively). For example, when men and women took a task described either as measuring the 'verbal skills of men and women' or simply measuring 'verbal abilities' the men changed their approach to the task depending on how the task was framed, that is, whether they were working under the threat of reinforcing the stereotype of male verbal inferiority.[26] Under stereotype threat, they tried harder to avoid doing badly (as opposed to trying to do well); they were slower and made fewer errors. The same researchers also showed the benefits for thinking of being positively, rather than negatively, stereotyped on a task. In the brick task, the participants have to think up as many creative uses for a brick as they can. Answers are rated for creativity, from original answers such as 'to show that I am just another brick in the wall' to distinctly less creative ones like 'to build a house'. Students told that people from their discipline tend to do very well on this task got significantly higher creativity ratings than students given the opposite stereotype about their discipline. It's not hard to see what a boost it could be in the real world, for prevailing cultural beliefs to push you towards a more open, imaginative, thinking style. In *Outliers*, Malcolm Gladwell compares two very high-IQ students' responses on the brick test. One student offered several creative examples (such as 'To use in smash-and-grab raids'). The other student, despite having an extraordinarily high IQ, came up with only two mundane ideas: 'Building things, throwing.' Gladwell rhetorically asks, 'Now which of these two students do you think is better suited to do the kind of brilliant, imaginative work that wins Nobel Prizes?'[27]

With horrible irony, the harder women try to succeed in quantitative domains, the greater the mental obstacles become, for several reasons. Stereotype threat hits hardest those who actually care about their maths skills and how they do on tests, and thus have the most to lose by doing badly, compared with women who don't much identify with maths.[28] Also, the more difficult and nonroutine the work, the more vulnerable its performance will be to the sapping of working memory, and possibly the switch to a more cautious problem-solving strategy.[29] There is also the problem that, as she proceeds up the career ladder, the mathematically minded woman will become increasingly outnumbered by men. In the United States, by 2001 women were earning about half of bachelor's degrees in mathematics, but only 29 percent of PhDs, and their numbers continue to diminish the further up the ladder you get.[30] This can compound her problem in more than one way. Her sex will become more and more salient, which in itself can trigger stereotype threat processes. One study even found that the more men there are taking a maths test in the same room as a solo woman, the lower women's performance becomes.[31] And, surrounded by men, she herself may come to grudgingly believe that women are indeed naturally inferior in maths – and women who endorse gender stereotypes about maths seem to be especially vulnerable to stereotype threat.[32]

But even in the absence of conscious endorsement of the stereotype, the maths-equals-male link will become ever more entrenched in her mind. With yet more dreadful irony, it may be women who are the most dedicated to maths who have the strongest maths-equals-male implicit associations. Amy Kiefer and Denise Sekaquaptewa at the University of Michigan used the Implicit

Association Test described earlier to see how strongly female college students implicitly associated maths with males. Overall, the women were quicker to pair words like *calculate*, *compute*, and *maths* with male words (like *he*, *him*, and *male*) than with female words. Interestingly though, the harder the maths class a woman had most recently taken, the stronger her maths-equals-male tendency. The researchers suggest that this is because harder classes are more male dominated, and so the link between maths and male gets reinforced in the mind. Unfortunately, women with especially strong maths-equals-male implicit associations seem to be at risk of being in a state of perpetual stereotype threat. Students with lower levels of implicit maths-equals-male associations showed the usual boost in performance when a hard maths test was presented in a nonthreatening way. But for women with very strong maths-equals-male associations, the dispersal of stereotype threat in the situation didn't help. The researchers suggest that this is because the stereotype is so firmly entrenched in the mind that it is resistant to alleviation.[33]

As our mathematical woman moves up the ranks, she will also progressively lose one very effective protection against stereotype threat: a female role model to look up to. People's self-evaluations, aspirations and performance are all enhanced by encountering the success of similar role models – and the more similar, the better.[34] In line with this, it's been found that the presence – real or symbolic – of a woman who excels in maths somehow serves to alleviate stereotype threat.[35] But of course the higher up the ladder a woman climbs, the harder she will have to look to find someone successful above her – either contemporary or historical – who is like her.

Finally, some intriguing research now hints that negative stereotypes about women may be particularly harmful to precisely the sort of woman who is disposed to struggle hardest to climb the career ladder. Some researchers speculate that higher testosterone levels are associated with a drive to gain and maintain status, in both men and women. Robert Josephs and his colleagues have been exploring the idea that high-T (high-testosterone, relative to others of the same sex) men and women are cognitively at their best when they are in situations that fit their testosterone-based drive to attain and maintain high status. By contrast, low status, or a threat to status, creates a mismatch for the high-T individual that has detrimental cognitive effects. (The basic theory behind this idea is that while the cognitive, emotional, and physiological reactions of the high-T person to a loss of status may be unproblematic when status can be restored by way of a fistfight, they are less helpful when status must be gained through a clever move of the bishop on the chessboard, a brilliant closing argument in court, or a publication in *Nature*.) In line with this idea, Josephs and his team found that high-T men and women, when put in a low-status position in the lab, underperformed on cognitive tests like the analytic and quantitative portions of the GRE and mental rotation.[36] By contrast, high testosterone works to their advantage when the situation yields an opportunity to enhance status. Josephs and colleagues found that high- and low-T men given a maths test described as identifying weak maths ability performed equivalently. But when the same test supposedly identified exceptional talent, the high-testosterone men rose to the challenge to enhance their maths status, and outperformed both the low-T and the other high-T men given the 'weak' maths test.

Although it may seem strange to think of women as being high T, it's important to bear in mind that when researchers measure testosterone levels in the saliva they are not directly indexing the amount of testosterone acting on the brain. All sorts of other factors are important, such as the number of receptors for that hormone in the brain, the sensitivity of those receptors, and the amount of bound versus free hormone in the blood (only free hormone molecules can bind to receptors).[37] It's even been suggested that women are more sensitive, neurally, to testosterone, or changes in its levels.[38] 'These complications raise the question of how we would measure the effective concentration of a sex hormone', points out University of New England neurobiologist Lesley Rogers.[39]

In any case, the interaction between testosterone level and status, and its effect on cognitive performance, seems to apply to women and men alike. But gender stereotypes add an extra layer of complexity to the situation. As Josephs and his colleagues point out, 'through its hierarchical ordering of two or more groups, a stereotype is essentially a statement about dominance or status.'[40] When the stereotype of women's inferiority in maths is made salient, a woman doing a maths test is at risk of confirming her lower status in the hierarchy of numeracy. Josephs and his colleagues predicted that because women with high testosterone are more concerned with their status, they will be particularly vulnerable to stereotype threat. In line with this, Josephs and his colleagues found that stereotype threat only impaired the performance of high-, but not low-, testosterone women. What does this imply for the world beyond the laboratory? It suggests that a talented, high-testosterone man is perfectly placed to rise to the challenge of opportunities that can enhance his status. Yet the situation is completely different for the equally talented high-T woman. Negative stereotypes about her group's ability create a cognition-impairing mismatch between her desire for high status and the low status that the stereotype ascribes to her. She's dancing backwards in high heels.

Imagine, just for a moment, that we could reverse the gender imbalance in maths and the maths-intensive sciences with a snap of our fingers, fill people's minds with assumptions and associations linking maths with natural female superiority, and then raise a generation of children in this topsy-turvy environment. Now it is males whose confidence is rattled, whose working memory resources are strained, whose mental strategies become nitpicky and defensive, and who look in vain for someone similar to inspire them. It's the boys in the classroom, not the girls, in whom researchers discover evidence that stereotype threat is already at work.[41] It is women who can now concentrate on the task with ease, whose alleged superiority brings creativity and boldness to their approach, who need only glance around the corridors of the department, the keynote speaker lineup, or the history books to see someone whose successes can seep into the very fabric of their own minds. What, we have to ask ourselves, would happen? Would male 'inherent' superiority reassert itself, would we quickly settle into some kind of equality, or – is it possible? – would the invisible hand of stereotype threat maintain the new status quo for decades to come?

The point of this armchair experiment is not to try to deny the many other factors that, no doubt, contribute in complex ways to gender inequality in scientific domains. But this body of research reminds us, again, that everything we do – be it maths, chess, child care, or driving – we do with a mind that is exquisitely sensitive to the social environment around it. Social psychologist Brian Nosek and his colleagues recently collected more than 500,000 scores from around the world on the gender-science Implicit Association Test (which measures how much easier it is to pair masculine words with science words and feminine words with liberal arts words, relative to the opposite female-science/male-arts pairing). They then compared this with data from the 2003 Trends in International Mathematics and Science Study (TIMSS) that measured maths and science achievement in eighth graders in thirty-four countries. Intriguingly, they found that across countries, over and above the effect of consciously reported stereotypes, the more strongly males are implicitly associated with science and females with liberal arts, the greater boys' advantage in science and maths in the eighth grade. (In some countries, it's worth saying, girls outscore boys.) Pointing out that 'social realities . . . shape minds', the researchers suggest that implicit gender stereotypes and the gender gap in science and maths achievement may be 'mutually reinforcing' – each feeding the other.[42]

The scalp serves as no barrier at all to the psychologically draining or boosting effects of pervasive cultural beliefs. And, as we'll see in the next chapter, social clues as to who belongs where also travel easily from environment to mind.

In the opening of her book *Brain Gender*, Cambridge University psychobiologist Melissa Hines dryly reports on the experience of being, in 1969, a member of the first freshman class at Princeton University to include women. Having been assigned by the university to what was described as a 'two-man room', she was allocated to a precept leader who 'called me Mr Hines for several weeks, apparently before realising that I was not male.'[1] A similar confusion over sex identity surrounded Sally Haslanger, now a philosophy professor at the Massachusetts Institute of Technology. When she received a distinction in her graduate exams, 'it seemed funny to everyone to suggest I should get a blood test to determine if I was really a woman.'[2]

Mary Beard, a classics professor at Cambridge University, recalls the Roman epigraphy classes she took as an undergraduate in the 1970s, 'where her tutor would pose "clever questions for the clever men and domestic questions for the dumb girls".'[3] At least there *were* questions for the 'girls'. Mary Mullarkey, who eventually became Chief Justice of the Colorado Supreme Court, was one of the few women to be enrolled at Harvard Law School in 1965. Although it had been fifteen years since the decision to admit women, she describes the change as still being, to many, 'a raw wound'. Mullarkey and her friend Pamela (Burgy) Minzer (destined to become Justice of the New Mexico Supreme Court), waited in vain to be called upon in their property class. Asking a woman to answer a question about law was an event considered by the professor of the class best limited to 'Ladies' Day'. The topic for that day, when it finally arrived, was marital gifts:

> Leaning over, [Professor] Casner said to me, 'Miss Mullarkey, if you were engaged – and I notice you're not' – he paused for laughter – 'would you have to return the ring if you broke the engagement?' That was the sole question asked of me in a full-year property class.[4]

Nor, Mullarkey and Burgy found, was a degree from Harvard Law School the same ticket to successful employment that it was for their male counterparts. Even though the federal Civil Rights Act, passed in 1964, prohibited employment discrimination based on gender, strangely, the law firms seemed unaware of the legal situation. 'It was commonplace for a law firm recruiter to tell a woman to her face that, although he would be willing to hire her, his senior partners or the firm's clients would never agree to have a female lawyer', Mullarkey recalls.[5]

It doesn't require any special sociological training to read the barely veiled message being communicated to these talented and ambitious women: *You don't belong here.* We tend to think of this sort of outright sex discrimination as being a thing of the past in Western, industrialised nations. *The Sexual Paradox* author Susan Pinker, for instance, writes of barriers to women as having been 'stripped away'.[6] Her book is peopled with women who, when asked if they've ever experienced ill-treatment because of their sex, scratch their heads and search the memory banks in vain for some anecdote that will show how they have had to struggle against the odds stacked against women. As we'll see in a later chapter, blatant, intentional discrimination against women is far from being something merely to be read about in history books. But here we're going to look at the subtle, off-putting, *you don't belong* messages that churn about in the privacy of one's own mind.

As we learned in the previous chapter, women who are invested in masculine domains often have to perform in the unpleasant and unrewarding atmosphere created by stereotype threat. Anxiety, depletion of working memory, lowered expectations, and frustration can all ensue. But there is a solution, albeit a rather radical one. As Claude Steele observed, 'women may reduce their stereotype threat substantially by moving across the hall from math to English class'.[7] Stereotype threat can do more than impair performance – it can also reduce interest in cross-gender activities.

A striking demonstration of this was provided by Mary Murphy and her colleagues at Stanford University. Advanced maths, science and engineering (MSE) majors were asked to give their opinion on an advertising video 'for an MSE summer leadership conference that Stanford was considering hosting the next summer'.[8] Under the cover story that the researchers were also interested in physiological reactions to the video, heart rate and skin conductance were recorded, to give a measure of arousal. After watching the ad, the students were asked questions to assess how much they felt they would belong at such a conference, and how interested they were in attending. There were two, near-identical videos, depicting about 150 people. However, in one video the ratio of men to women approximated the actual gender ratio of MSE degrees: there were three men to every woman. In the second video, men and women were featured in equal numbers. Women who saw the gender-equal video responded very much like men, both physiologically and in their sense of belonging and interest in the conference. But for women who saw the more realistically imbalanced version, it was a very different experience. They became more aroused – an indicator of physiological vigilance. They expressed less interest in attending the conference when it was gender unbalanced. (Interestingly, so did men – although this was probably, one can't help but think, for different reasons.) And although women and men who saw the gender-balanced video very strongly agreed that they belonged there, the conviction of this agreement among women who saw a gender imbalance was significantly lower. Under the naturalistic condition of male dominance, they were no longer so sure that they belonged.

Being outnumbered by men is simply a fact of life for women in MSE domains – as is being exposed to gender stereotypes in advertising. At first, it's not obvious why an advertisement depicting, say, a woman bouncing on her bed in rapture over a new acne product might serve as a psychic obstacle to women looking to enter masculine fields. However, images of women fretting over their appearance or in ecstasy over a brownie mix, although they have nothing to do with mathematical ability directly, nonetheless make gender stereotypes in general more accessible. Paul Davies and his colleagues showed either these or neutral commercials to women and men who were invested in doing well in maths. They were then given a GRE-like exam that had both maths and verbal problems. Men in both conditions, and women who had seen neutral ads, attempted more maths problems than verbal ones. But women who had seen the sexist ads showed exactly the opposite pattern, avoiding the maths questions. Their career aspirations were also influenced, with a flipping of occupational preferences, from those that require strong mathematical skills (like engineer, mathematician, computer scientist, physicist and so on) to those that depend more heavily on verbal abilities (such as author, linguist and journalist).[9] Ads that trade in ditzy stereotypes of women also, Davies and colleagues found, reduce women's interest in taking on a leadership role. Male and female university students were equally interested in leading a group – except for women exposed to the gender-stereotyped commercials, who were more likely to choose a nonleadership role instead.[10]

Entrepreneurship is another male-dominated arena, and one in which the traits usually assumed to be vital for success – strong-willed, resolute, aggressive, risk-taking – have a decidedly male feel. Here, then, is another occupational niche to which women could easily be made to feel that they don't belong. Female business school students were given one of two fabricated newspaper articles to read. One described entrepreneurs as creative, well-informed, steady and generous – and claimed that these qualities are shared equally between men and women. The other article, however, depicted the prototypical entrepreneur as aggressive, risk-taking and autonomous, all traits that belong firmly in the male stereotype. The women were then asked how interested they were in being self-employed, and owning a small or high-growth business. For women who scored low on a proactive measure (the tendency to 'show initiative, identify opportunities, act on them, and persevere until they meet their objectives') it made no difference which article they read. But what about the highly proactive women? As you might expect of these go-getting women, their interest in an entrepreneurial career was high but significantly reduced after reading the entrepreneurship-equals-male news article.[11]

What psychological processes lie behind this turning away from masculine interests? One possibility is that, as we learned in an earlier chapter, when stereotypes of women become salient, women tend to incorporate those stereotypical traits into their current self-perception. They may then find it harder to imagine themselves as, say, a mechanical engineer. The belief that one will be able to fit in, to belong, may be more important than we realise – and may help to explain why some traditionally male occupations have been more readily entered by women than others.[12] After all, the stereotype of a vet is not the same as that of an orthopaedic surgeon or a computer scientist, and these are different again from the stereotype of a builder or a lawyer. These different stereotypes may be more or less easily reconciled with a female identity. What, for example, springs to mind when you think of a computer scientist? A man, of course, but not just any man. You're probably thinking of the sort of man who would not be an asset at a tea party. The sort of man who leaves a trail of soft-drink cans, junk-food wrappers, and tech magazines behind him as he makes his way to the sofa to watch *Star Trek* for the hundredth time. The sort of man whose pale complexion hints alarmingly of vitamin D deficiency. The sort of man, in short, who is a geek.

Sapna Cheryan, a psychologist at Washington University, was interested in whether the geek image of computer science plays a role in putting off women. When she and her colleagues surveyed undergraduates about their interest in being a computer science major, they found, perhaps unsurprisingly given that computer science is male-dominated, that women were significantly less interested. Less obvious, however, was *why* they were less interested. Women felt that they were less similar to the typical computer science major. This influenced their sense that they belonged in computer science – again lower in women – and it was this lack of fit that drove their lack of interest in a computer science major.[13]

However, an interest in *Star Trek* and an antisocial lifestyle may not, in fact, be unassailable correlates of talent in computer programming. Indeed, in its early days, computer programming was a job done principally by women and was regarded as an activity to which feminine talents were particularly well suited. 'Programming requires lots of patience, persistence and a capacity for detail and those are traits that many girls have' wrote one author of a career guide to computer programming in 1967.[14] Women made many significant contributions to computer science development and, as one expert puts it, '[t]oday's achievements in software are built on the shoulders of the first pioneering women programmers.'[15] Cheryan suggests that '[i]t was not until the 1980s that individual heroes in computer science, such as Bill Gates and Steve Jobs came to the scene, and the term "geek" became associated with being technically minded. Movies such as *Revenge of the Nerds* and *Real Genius*, released during these years, crystallized the image of the "computer geek" in the cultural consciousness.'[16]

If it is the geeky stereotype that is so off-putting to women, then a little repackaging of the field might be an effective way of drawing more women in. Cheryan and her colleagues tested this very idea. They recruited undergraduates to participate in a study by the 'Career Development Center regarding interest in technical jobs and internships.' The students filled out a questionnaire about their interest in computer science in a small classroom within the

William Gates Building (which, as you will have guessed, houses the computer science department). The room, however, was set up in one of two ways for the unsuspecting participant. In one condition, the décor was what we might call geek chic: a *Star Trek* poster, geeky comics, video game boxes, junk food, electronic equipment and technical books and magazines. The second arrangement was substantially less geeky: the poster was an art one, water bottles replaced the junk food, the magazines were general interest and the computer books were aimed at a more general level. In the geeky room, men considered themselves significantly more interested in computer science than did women. But when the geek factor was removed from the surroundings, women showed equal interest to men. It seemed that a greater sense of belonging brought about this positive change. Simply by altering the décor, Cheryan and colleagues were also able to increase women's interest in, for example, joining a hypothetical Web-design company. The researchers note 'the power of environments to signal to people whether or not they should enter a domain', and suggest that changing the computer science environment 'can therefore inspire those who previously had little or no interest . . . to express a new-found interest in it.'[17]

You might think that this is a nice sentiment, but that a narrowly focused, unsociable personality simply goes hand-in-hand with talent in computer science. But as developmental psychologist Elizabeth Spelke and Ariel Grace point out, 'personality traits that are *typical* of a given profession often are mistakenly thought to be *necessary* to the practice of the profession.' They provide, as a historical example, the assumption by an early-twentieth-century psychologist that his talented Jewish students could not succeed in academia because they did not share the traits of the predominantly Christian faculty: he 'mistakenly assumed that the typical mannerisms of his Harvard colleagues were necessary for success in science.'[18]

Underscoring Spelke and Grace's point is a fascinating natural experiment in the Carnegie-Mellon computer science department that suggests that geeky traits may indeed be extrinsic to success in computer science. In the mid- to late 1990s, an intensive study of male and (the very few) female computer science students at Carnegie-Mellon found that the men were very focused on programming – the sort of person who 'dreams in code' – while the few women in the programme were more interested in the applications of computer science. But in the late 1990s, the admission criteria were changed so as to no longer unnecessarily and unfairly exclude applicants without a lot of programming experience.[19] This led to a fivefold increase in the number of women, from about 7 percent to 34 percent. Lenore Blum and Carol Frieze took the opportunity of this situation to interview the students who entered the computer science programme in 1998. In 2002, when they were interviewed, these students were, uniquely, the babies of the old, hacker-favouring admission criteria, yet were now in a department with a much more diverse student body. Remarkably, Blum and Frieze found that interest in programming versus applications was now a point of similarity, rather than difference, between men and women. 'Almost all students saw programming as one part of their interests and the computers as a "tool" for their primary focus, which was applications.' But also, there was evidence that the 'students were constructing a new image', and one in which the 'narrowly focused computer science student' was no longer the norm:

> Our cohort included students who played the violin, wrote fiction, sang in a rock band, participated in university team sports, enjoyed the arts, and were members of a wide range of campus organizations. We found that men and women alike appear to be moving towards a more well-rounded identity that embraced academic interests and a life outside of computing. Students described themselves as 'individual and creative, just interesting all-round people', 'very intelligent, . . . very grounded, not the traditional geek . . .', 'much more well rounded than people five or six years ago.'

Recall that these students had been chosen according to the old criteria. They were the geeky programmers. And yet, as the researchers suggest, the years spent in an increasingly gender-equal environment 'had shaped their image of themselves. We might also speculate that such a transitional culture gave the men "permission" to explore their nongeeky characteristics'.[20]

Both women and computer science are the losers when a geeky stereotype serves as an unnecessary gatekeeper to the profession. And recent work by psychologist Catherine Good and her colleagues shows that a 'sense of belonging' is also an important factor in women's intention to continue in maths. This feeling of belonging, however, can be eroded by an environment that communicates that maths ability is a fixed trait and not something that hard work can increase, especially in combination with the message that women are naturally less talented than men, Good and colleagues found.[21] Philosopher Sally Haslanger has suggested that a difficulty even today for women (and minority) philosophers is that 'it is very hard to find a place in philosophy that isn't actively hostile towards women and minorities, or at least assumes that a successful philosopher should look and act like a (traditional, white) man.'[22]

But choosing a career is not just about finding a place socially in which one can feel at home. It also entails finding a fit with one's talents. People of course tend to be drawn towards jobs in which they are likely to succeed. If gender stereotypes can affect people's perceptions of their abilities (as we now know that they can), then it would not be surprising to discover that this then has effects on career decisions. Sociologist Shelley Correll has shown that beliefs about gender differences in ability have an important role to play in people's perceptions of their own masculine abilities and, as you might expect, this affects their interest in careers that rely on such skills. Correll used the data from the 1988 National Educational Longitudinal Study, involving tens of thousands of high school students, to carefully compare students' actual grades with their own assessments of their mathematical and verbal competence. She found that boys rated their maths skills higher than

their equal female counterparts. This was likely due to the culturally shared belief that males *are* better at maths, because boys were
selective in their self-embellishment: they didn't inflate their verbal competence. These self-assessments proved to be an important
factor in the students' decision making about their careers. With
actual ability (assessed by test scores) held equal, the higher a boy
or girl rates his or her mathematical competence, the more likely it
is that he or she will head down a path towards a career in science,
maths or engineering. Correll concludes that 'boys do not pursue
mathematical activities at a higher rate than girls do because they
are better at mathematics. They do so, at least partially, because
they *think* they are better.'[23] For example, gender differences in
self-assessment of maths ability fully explained the gender gap in
calculus enrolments.

Correll then went on to show just how easy it is to create a
gender stereotype that diminishes women's confidence and interest
in a supposedly male domain. She used a contrast sensitivity test,
in which the participant has to guess which colour, black or white,
covers a greater area in a series of rectangles. Her participants,
freshmen at Cornell University, were told that 'a national testing
organisation developed the contrast sensitivity exam and that both
graduate schools and Fortune 500 companies have expressed interest in using this exam as a screening device.'[24] (In truth, the test is
a fake one: black and white appear in essentially equal proportions,
so there is actually no correct solution.) Participants were then told
either that males, on average, perform better on tests of contrast
sensitivity or that there is no gender difference.

The participants were all given the same feedback on their
test performance, but *how* this score was perceived depended on
the context – male-advantage or gender-equal – in which the test
was presented. When the students thought that contrast sensitivity was a nongendered ability, women and men's self-assessments
were very similar. But it was a different story when the underlying
assumption was that one sex had the upper hand. In this male-
favourable context, men rated their contrast sensitivity ability more

highly and claimed to have done better on the tasks. They also set themselves a more lenient standard against which to judge their performance. Correll then investigated whether, as in her real-world data set, higher self-assessments would lead to higher aspirations. She found that they did. When men thought that they were, as a group, better at contrast sensitivity, they were more likely than women to say that they would enrol for courses or seminars based on the ability, and to apply for graduate programmes or high-paying jobs that relied heavily on the skill. And it was their higher self-assessments of ability that appeared to bring about this greater interest in contrast-sensitivity-based aspirations. We like what (we think) we are good at.

But of course many women do persist in male-dominated careers like mathematics, despite the stereotype threat and lack of sense of belonging. Luckily for them, there is an alternative to turning away from maths – and this is to turn away from being female. Emily Pronin and her colleagues found that female undergraduates at Stanford University who had taken more than ten quantitative courses were less likely than other women to rate as important and applicable to them supposedly maths-incompatible behaviours such as wearing makeup, being emotional, and wanting children.[25] The researchers then went on to provide evidence that it is not simply that women who like to wear lipstick and fondly imagine having children one day are *intrinsically* less interested in maths. Rather, women who want to succeed in these domains strategically shed these desires in response to reminders that maths is not for women. The researchers recruited a group of Stanford undergraduate women, for all of whom maths ability was important. Half of the women read a (fabricated) scientific article about ageing and verbal ability. But the remainder of the women read a shortened version of an actual scientific article about gender and maths, published in *Science*.[26] This was a study of the Scholastic Aptitude Test results in maths for nearly 10,000 high-achieving seventh and eighth graders. Boys were more likely to score highly than girls, and the article concluded that there is 'a substantial sex difference in mathematical reasoning ability in favour of boys',[27] together with the assertion that this advantage reflects boys' innate superiority in spatial ability.

The women certainly found the article threatening, and put some effort towards challenging its findings and conclusions. But it still had an effect on them. Women who read the nonthreatening article identified equally with feminine characteristics believed to be both relevant and irrelevant to maths-related careers. But the women who had read the *Science* article about maths and gender identified less with female characteristics regarded as a liability in quantitative domains. Parts of their identity were being hurled overboard in an attempt to remain afloat in male-dominated waters. If these are particularly cherished parts of the self-concept that must be abandoned then, in the end, the woman may prefer for the boat to sink.

The behaviour of colleagues may also sometimes make it harder to keep female and work identities compatible in male-dominated domains. The recent Athena Factor report conducted by the Center for Work-Life Policy found that a quarter of women in corporate engineering and technology jobs thought that their colleagues believed their sex to be intrinsically inferior in scientific aptitude. '[M]y opinions and reasoning are always questioned, "Are you sure about that?"' complained one focus group participant, 'whereas what the men say is taken as gospel.' The focus groups of the Athena report told tale after tale with a common theme: female engineers whom men assumed were administrative assistants; senior women assumed to be the most junior person in the room; double takes in the meeting room at the sight of a woman.[28] In reaction to the Athena report, a woman in a senior engineering position blogged that '[m]any of our clients think I'm in the meetings to take notes for the men . . . some even apologise for boring me with the technical discussions, assuming I have no idea what they're talking about.'[29] It's not hard to see that these sorts of attitudes and assumptions could not only rapidly become rather tiresome but also chip away at women's sense of belonging.

Echoing Emily Pronin and her colleagues' discovery that mathematically inclined women shed the feminine attributes they
perceived as a liability, the Athena report sketches a disquieting
picture of the psychological changes that take place in women who
remain in SET careers. For the easiest solution to the problem of
being female in a setting in which women are made to feel that
they are inferior and do not belong is to become as unfeminine as
possible. At the most superficial level, makeup, jewellery and skirts
– icons of femininity that draw attention to their wearer's femininity – were rarely in evidence, the researchers noted. The women
also took up antifemale attitudes, denigrating other women as
emotional, and 'heaped scorn' on women-focused programmes and
any work-related gatherings dominated by women. 'By definition
nothing important is going on in this room: In this company men
hold the power', was how one female engineer explained her policy
of avoiding female work gatherings. The awful, intractable incom-
patibility of being a woman in a male-dominated SET workplace
was starkly encapsulated by one woman quoted in the report who
described how, more and more, she had developed a 'discomfort
with being a woman.'[30]

As the arguments that women lack the necessary intrinsic talent
to succeed in male-dominated occupations become less and less
convincing, the argument that women are just less interested has
grown and flourished.[31] Yet as we've seen in this chapter, interest
is not impervious to outside influence, at least in the young adult
samples with which most of this research is done. It is remarkably
easy to adjust the shine of a career path for one sex. A few words
to the effect that a Y chromosome will serve in your favour, or
a sprucing up of the interior design, is all that it takes to bring
about surprisingly substantial changes in career interest. Having
seen what effect on career interests a simple, brief manipulation in
the lab can have, one can't help but wonder at the cumulative influence of that giant, inescapable social psychology lab known as life.

The existing gender inequality of occupations, the sexist ads, the opinions of presidents of high-profile universities, not to mention all the 'brain facts' that we'll get to later – these all interact with, and shape, our minds.

And then, there are people in our lives whose minds, just like ours, are richly endowed with implicit and explicit attitudes about gender. The tilting of the playing field that their half-changed minds and behaviour create, as we'll see in the final few chapters of this part of the book, are still an important part of the half-changed world.

In her book *Scientists Anonymous*, Patricia Fara describes how, around the turn of the nineteenth century, botanist Jeanne Baret and mathematician Sophie Germain were obliged to present themselves as men to carry out their research.[1] Unlike Baret, today's female biologists do not have to pretend to be men to carry out fieldwork. Nor do contemporary female mathematicians need to employ Germain's subterfuge, studying by correspondence under cover of a male identity. Yet even today, the evidence suggests that it would be a shrewd career move for a woman to disguise herself as a man. People who *have* transformed their identity in this way – namely, female-to-male transsexuals – report decidedly beneficial consequences in the workplace. Ben Barres is a professor of neurobiology at Stanford University, and a female-to-male transsexual. In an article in *Nature* he recalls that '[s]hortly after I changed sex, a faculty member was heard to say "Ben Barres gave a great seminar today, but then his work is much better than his sister's."'[2] Similar stories cropped up in a recent interview study of twenty-nine female-to-male transsexuals. Kirsten Schilt, a Research Fellow at Houston's Rice University, interviewed the men about their work experiences both before and after their transition from women to men. Her study reveals that many immediately enjoyed greater recognition and respect. Thomas, an attorney, related how a colleague praised the boss for getting rid of Susan, whom he regarded as incompetent. He then added that the 'new guy', Thomas, was 'just delightful' – not realising, of course, that Thomas and Susan were one and the same. Roger, in retail, found that now that he is a man people bypass his female boss and beeline straight to him with their questions. Paul, continuing his work in secondary education, suddenly found himself being continually called upon in meetings to offer his newly valuable opinions. And several blue-collar workers reported that work is a great deal easier since transition.[3]

As Barres rightly acknowledges, anecdotes are not data. But these insights from the experiences of people who have lived on both sides of the gender divide offer an intriguing glimpse into the possibility that a person's talents in the workplace are easier to recognise when that person is male. Empirical research points to the same conclusion.

First, there are experimental studies showing that men's qualifications, talents, and achievements shine brighter and provide a better fit with the demands of a nonfeminine job – even when identical to those of a woman.[4] For example, in one recent study more than 100 university psychologists were asked to rate the CVs of Dr. Karen Miller or Dr. Brian Miller, fictitious applicants for an academic tenure-track job. The CVs were identical, apart from the name. Yet strangely, the male Dr. Miller was perceived (by both male and female reviewers) to have better research, teaching and service experience than the luckless female Dr. Miller. Overall, about three-quarters of the psychologists thought that Dr. Brian was hireable, while only just under half had the same confidence in Dr. Karen.[5] The same researchers also sent out applications for the position of tenured professor, again identical but for the male and female name at the top. This time, the application was so strong that most of the raters thought that tenure was deserved, regardless of sex. However, the endorsement of Karen's application was four times more likely to be accompanied by cautionary caveats scrawled in the margins of the questionnaire: such as, 'I would need to see evidence that she had gotten these grants and publications on her own' and 'We would have to see her job talk'.[6]

A meta-analysis of the employment prospects of so-called paper people (fictitious job applicants evaluated in the lab) found that, overall, men are indeed rated more favourably than identical women for masculine jobs (while participants are biased against paper men applying for stereotypically feminine jobs, like secretarial work or teaching home economics).[7] What is the problem for women seeking a job outside the 'pink ghettos' of secretarial work, teaching and health? One possibility is the 'lack of fit' between the communal stereotype of women and demanding professional roles. As one of the leading researchers in this area, New York University's Madeline Heilman, has explained:

> Essential to understanding how the female gender stereotype can obstruct women from advancing up the organizational hierarchy is the realization that top management and executive level jobs are almost always considered to be 'male' in sex-type. They are thought to require an achievement-oriented aggressiveness and an emotional toughness that is distinctly male in character and antithetical to both the stereotyped view of what women are like and the stereotype-based norms specifying how they should behave.[8]

In other words, both the *descriptive* ('women are gentle') and the *prescriptive* ('women *should be* gentle') elements of gender stereotypes create a problem for ambitious women. Without any intention of bias, once we have categorised someone as male or female, activated gender stereotypes can then colour our perception. When the qualifications for the job include stereotypically male qualities, this will serve to disadvantage women (and vice versa). In one classic study, Monica Biernat and Diane Kobrynowicz gave undergraduates a job description and a candidate résumé.[9] For every participant, the job description was identical except for the job title: either executive secretary or executive chief of staff. (Of course, the latter was intended to come across as more masculine, higher status and better paid.) Each participant also received the same résumé, except some evaluated Kenneth Anderson and the remainder evaluated Katherine Anderson. These evaluations revealed a favouring of, and greater confidence in, female secretaries and male chiefs of staff.[10]

The lack-of-fit bias may act particularly strongly against mothers. Using the cover story that a start-up communications company was looking for a head for its marketing department, sociologist Shelley Correll and colleagues found that, compared with paper nonmothers, identical paper mother applicants were rated about 10 percent less competent, 15 percent less committed to the workplace and worthy of $11,000 less salary. Moreover, only 47 percent of mothers, compared with 84 percent of nonmothers were recommended for hire.[11] One only hopes that the little paper children are worth the career sacrifice. As a follow-up, over the course of eighteen months Correll and her colleagues sent out a total of 1,276 fictitious résumés and cover letters for real marketing and business jobs advertised in the press. Each employer was sent two applications from two equally qualified applicants. They were both the same sex (sometimes both male, other times both female), but only one was identifiable as a parent. (The researchers counterbalanced which applicant was the parent.) Then the researchers sat back and waited to see who got the most callbacks from the potential employers. While parenthood served as no disadvantage at all to men, there was evidence of a substantial 'motherhood penalty'. Mothers received only half as many callbacks as their identically qualified childless counterparts. Ongoing research is investigating whether these days it is especially mothers who are discriminated against.[12]

While stereotypes can distort our perception of others, they are not so powerful that they can blind us to actual evidence that a female candidate has the necessary confidence, independence and ambition to succeed in leadership roles. Now, however, the female candidate comes up against the prescriptive part of the gender stereotype:

There is no form of human excellence before which we bow
with profounder deference than that which appears in a delicate woman, adorned with the inward graces and devoted to
the peculiar duties of her sex; and there is no deformity of
human character from which we turn with deeper loathing
than from a woman forgetful of her nature, and clamorous for
the vocation and rights of men.[13]

Though professor of mathematics, lawyer, and political writer
A. T. Bledsoe uttered the words above in 1856, there is still a resid-
ual unease – both conscious and implicit – with women in positions of power.[14] When women display the necessary confidence
in their skills and comfort with power, they run the risk of being
regarded as 'competent but cold': the bitch, the ice queen, the iron
maiden, the ballbuster, the battle axe, the dragon lady . . . The
sheer number of synonyms is telling. Put bluntly, we don't like the
look of self-promotion and power on a woman. In experimental
studies, women who behave in an agentic fashion experience backlash: they are rated as less socially skilled, and thus less hireable for
jobs that require people skills as well as competence than are men
who behave in an identical fashion. And yet if women don't show
confidence, ambition and competitiveness then evaluators may use
gender stereotypes to fill in the gaps, and assume that these are
important qualities she lacks. Thus, the alternative to being competent but cold is to be regarded as 'nice but incompetent'.[15] This
catch-22 positions women who seek leadership roles on a 'tightrope of impression management'.[16] In February 2006, the chair-
man of the Republican National Committee claimed that Hillary
Clinton was too angry to be elected president. As Maureen Dowd
noted in the *New York Times*, the 'gambit handcuffs Hillary: If
she doesn't speak out strongly against President Bush, she's timid
and girlie. If she does, she's a witch and a shrew.' In an empirical investigation of this damned-if-you-do, damned-if-you-don't
situation for women leaders, Victoria Brescoll and Eric Uhlmann
found that while expressing anger often enhances men's status and

competency in the eyes of others, it can be very costly to women in terms of how they are perceived.[17]

Motherhood, by the way, serves to upset an already delicate balance. Students rated a childless professional woman as more competent than warm, but an identical working mother as more warm than competent. Perhaps unsurprisingly, the working mother was thus also regarded as less valuable, less likely to be promoted and less worthy of training.[18] Suspiciously, this penalising of working mothers was justified by some as being 'because she telecommutes', even though telecommuting was of no concern whatsoever when performed by childless women and men, or fathers.

Rutgers University psychologist Laurie Rudman and her colleagues have recently discovered that what people find particularly objectionable in professional women are status enhancing behaviours like being aggressive, dominating and intimidating. For instance, in one study students read a letter of recommendation for an academic applying for promotion to English professor.[19] The fictional candidate was superb, an internationally renowned and highly intelligent author and literary critic. To this information it was added either that the applicant's style of literary criticism was tactful or ruthless. And, as you have already guessed, in one version of the letter the applicant was female (Dr. Emily Mullen) and, in the other, male (Edward). The tactful versions of Emily and Edward were equally well liked and rated equally hireable. However, the ruthless version of Edward was considered significantly more likeable and hireable than his female counterpart. The pitiless Emily was less hireable because she was disliked, and she was disliked because she was seen as more intimidating, dominant and ruthless than the identical Edward.

Of course, student participants in a laboratory experiment know that their humble opinions of candidates have no consequence for that person's career. They also do not face the prospect of having to confess, under probing by a hiring committee, that 'I just didn't like her.' But other lab experiments show how a man can be rated as more suitable for a masculine job simply by virtue of his maleness, but in apparently legitimate fashion. If, for example, you were recruiting for the position of manager in a construction company, what would you think was more important: experience or education? Michael Norton and his colleagues made up applications in which one of the two strongest candidates had better educational qualifications but less industry experience, while the other strong candidate had experience but a less impressive educational background. When the sex of the participant wasn't mentioned (probably most people assumed that both applicants were male), 76 percent of male undergraduates strongly preferred a better-educated candidate over one with more industry experience. Likewise, three-quarters of participants preferred a better-educated *male* candidate over a *female* candidate with more industry experience. In a fair and equal world, then, the better-educated female candidate would enjoy the same advantage over a lesser-educated, highly experienced male competitor. But she doesn't. Only 43 percent chose her. But this wasn't prejudice, you understand; or, at least, not of the conscious variety. After they ranked the candidates, the participants were asked to write down why they made the choice they did and the most important factor in their decision. Education was considered far more important when possessed to a greater degree by a male, rather than female, candidate. Yet even though gender clearly *was* influencing the evaluations, almost none of the participants mentioned it as a factor in their decision making.[20]

In a similar study conducted at Yale University, undergraduate participants were offered the opportunity to use the same kind of casuistry to maintain the occupational status quo. The students evaluated one of two applicants (Michael or Michelle) for the position of police chief. One applicant was streetwise, a tough risk-taker, popular with other officers, but poorly educated. By contrast, the educated applicant was well schooled, media savvy, and family oriented, but lacked street experience and was less popular with the other officers. The undergraduate participants judged the job applicant on various streetwise and education criteria, and then rated the importance of each criterion for success as a police chief.

Participants who rated Michael inflated the importance of being an educated, media-savvy family man when these were qualities Michael possessed, but devalued these qualities when he happened to lack them. No such helpful shifting of criteria took place for Michelle. As a consequence, regardless of whether he was streetwise or educated, the demands of the social world were shaped to ensure that Michael had more of what it took to be a successful police chief. As the authors put it, participants may have 'felt that they had chosen the right man for the job, when in fact they had chosen the right job criteria for the man.'[21] Ironically, the people who were most convinced of their own objectivity discriminated the most. Although self-reported endorsement of sexist attitudes didn't predict hiring bias, self-reported objectivity in decision making did.

This is unintended sex discrimination at work. Rather than unfairly stereotyping the candidates – assuming, for example, that Michael was tougher than Michelle – the raters instead 'defined their notion of "what it takes" to do the job well in a manner tailored to the idiosyncratic credentials of the person they wanted to hire'.[22] Recently, Laurie Rudman and her colleagues have shown that these 'shifting criteria' can be used to implement backlash against agentic women. Student participants watched a videotape of an interview for a computer lab manager position in which the applicant, either female or male, espoused either an agentic managerial style or a communal one. The agentic managers, for example, said things like, 'There's no question about it, I like to be the boss . . . I like being in charge – to be the person who makes the decisions'. As in other studies, the male agentic manager was rated as being both more socially skilled and more hireable than the female version. But this hiring discrimination was cleverly done. Participants weighted competence more heavily than social skills in their assessments – but with one exception. For female agentic managers alone, social skill score was more important than competence ratings. As the researchers point out, this strategy puts agentic women at a double disadvantage. Not only is their high competence discounted, but emphasised instead are the social skills that, you will recall, were rated unfairly low.[23]

Many, although not all, studies of real employment contexts also find that men are preferred over women for traditionally masculine positions – but both positive and negative findings from such studies are hard to interpret. The beauty of well-controlled experimental lab work is that you can, with absolute certainty, pinpoint sex discrimination. When Karen, Katherine, Michelle, and Emily are identical to Brian, Kenneth, Michael, and Edward there is little wiggle room for justifying differential treatment. The limitation of this kind of experimental work, however, is that it generally involves university students evaluating paper people. Real employers interviewing real people for real jobs will certainly be more motivated (as well as better qualified) to get the right person, as well as, sometimes at least, being more accountable for their decisions. This should count in favour of better and fairer decision making. But at the same time, one does not receive a new, more objective, mind upon graduation. Today's students are tomorrow's employers, and in the messier environment of real-world decision making there is ample scope for hiring criteria to shift, especially further up the career ladder where qualifications and experience become more idiosyncratic and harder to compare across candidates. So it's interesting – and entirely consistent with the research presented here – that University of California–Irvine maths professor Alice Silverberg has 'seen a variety of excuses used to justify not choosing a woman, which [she's] never seen used against a man'.[24]

The prescriptions of the communal stereotype can of course continue to disadvantage women even once they are hired. Unlike men in the same position, women leaders have to continue to walk the fine line between appearing incompetent and nice and competent but cold. Experimental studies find that, unlike men, when they try to negotiate greater compensation they are disliked. When they try out intimidation tactics they are disliked. When they succeed in a male occupation they are disliked. When they fail to perform the altruistic acts that are optional for men, they are disliked. When they *do* go beyond the call of duty they are not, as men are, liked more for it. When they criticise, they are disparaged. Even when they merely offer an opinion, people look displeased.[25] The perceptive reader will notice a certain pattern emerging. The same behaviour that enhances *his* status simply makes *her* less popular. It's not hard to see that this makes the goal of getting ahead in the workplace distinctly more challenging for a woman. This perceived dislikeability often drives economic and promotional penalties. And while not all occupations are justly described as a popularity contest, it is simply human nature to prefer to work with, and be around, someone you like. As Heilman points out:

> Upper management is sometimes referred to colloquially as a 'club.' Members of such clubs are apt to blackball the entry of those who seem inappropriate or distasteful. Simply put, if a woman is perceived as equally competent to a male colleague but seen as less interpersonally appealing and suitable as a member of the upper management team, there are likely to be unfavorable consequences for her in terms of rewards and advancement.[26]

All of which means that at a day-to-day level, women leaders may be in the tiresome double bind of directing, commanding and controlling their teams without appearing to do so. Deborah Cameron, discussing the work of Janet Holmes who recorded and analysed about 2,500 workplace interactions, describes how Clara, the team leader in a multinational company, uses a typically masculine style of leadership. It's firm, abrupt and direct. So, to deal with being issued orders by her, the team has developed a running joke whereby she is referred to as Queen Clara. For instance, when Clara says 'it's a no', one of her team members responds that it's a 'royal no'. As Cameron points out:

[W]ould a man in Clara's position who behaved in a similar way have to make the same concessions? Would he be
dubbed 'the King' by his subordinates, and teased about his
'royal' manner? Arguably, the humorous 'Queen Clara' per-
sona is needed to render Clara's style of management acceptable precisely because she is not a man. A woman who displays
authority as unabashedly as Clara still makes a lot of people
feel uncomfortable or threatened.[27]

At the end of the tightrope of impression management, should
it be successfully navigated, is the glass cliff. Michelle Ryan and
her colleagues noticed a curious pattern when they looked at the
share-price performance of the top 100 companies in the UK, both
before and after the appointment of male and female board members. In the months before a man was appointed to the board of
directors, company performance was relatively stable. But women
tended to be appointed after a period of consistently low performance. In other words, women were being appointed to positions
'associated with a higher risk of failure, and [that] were therefore
more precarious.'[28] Ryan and colleagues' follow-up studies back up
these data from real companies. Who do people choose to become
financial director of a company with declining share prices, to be
the lead lawyer for a case that is doomed to fail, to be the youth
representative for a failing music festival or to run for an unwin-
nable political seat? Students and senior business leaders choose
women for these risky, or simply dead-end, positions.[29]

Men aren't always the winners; the lack-of-fit phenomenon
can work against them, too. For example, when people were evaluating candidates for a position as women's studies professor, the
criteria (activist versus academic) were shifted to make the woman
the better candidate.[30] But often, when men choose to enter less-
prestigious female professions they quickly find rolled out for them
a red carpet leading to a better-paying position within the field.
The sociologist Christine Williams coined the term 'glass escalator' to encapsulate her discovery that men in (what are currently)

traditionally female occupations like nursing, librarianship and teaching 'face invisible pressures to move up in their professions. As if on a moving escalator, they must work to stay in place.'[31] (Recent research suggests that only white men can ride the glass escalator.)[32] Many of the men she interviewed suggested that there was a hiring preference for men and reported being 'kicked upstairs' into more masculine specialties, like administration, that also happened to be better-paying, higher-status positions. Sometimes it was actually a struggle for men to stay in the more-feminine roles that they preferred, so powerful were the assumptions of those around them that they should be somewhere else. Perceived as, in a sense, too competent for feminine occupations, they were tracked into more supposedly legitimate, prestigious ones.

The unwitting sex discrimination that devalues women's achievements and sets difficult standards for interpersonal behaviour perhaps explains why, in survey after survey, women consistently and reliably rate their jobs as simply harder work than do men. Using large data sets from both the United States and Britain, sociologists Elizabeth Gorman and Julie Kmec found that '[e]ven when women and men are matched on extensive measures of job characteristics, family and household responsibilities, and individual qualifications, women report that their jobs require more effort than men do.'[33] (As a former female investment banker recently commented in *The Observer*, 'We knew we had to work harder and be better than everyone else. The trading floor would empty out and after 7pm or 8pm only the women would be left. We would joke that we were doing our "vagina tax" work.')[34] Unconscious bias may also explain, in part, why women are paid less for the same work. As one comprehensive review of the literature concluded, 'Women earn less than men, and no matter how extensively regressions control for market characteristics, working conditions, individual characteristics, children, housework time, and observed productivity, an unexplained gender pay gap remains for all but the most inexperienced of workers.'[35] Interestingly, the implicit idea that a man's work is worth more than a woman's seems to be learned young. When eleven- to twelve-year-old children are shown pictures of men and women performing unfamiliar jobs, they rate as more difficult, better paid and more important those occupations that happen to be performed by men.[36]

We can be prejudiced even when we don't intend to be. Not many people would, I think, agree that women should be judged to a higher, harder, shifting standard; suggest that they be sanctioned for behaviour that is acceptable in men; or think it fair that they be paid less for the same work. But when we categorise someone as male or female, as we inevitably do, gender associations are automatically activated and we perceive them through the filter of cultural beliefs and norms. This is sexism gone underground – unconscious and unintended – and social psychologists and lawyers are becoming very interested in how this new, covert and unintended form of sexism disadvantages women (as well as non-whites) in the workplace. There's little doubt that this new form of subtle discrimination *is* important and *does* hold women back, especially, perhaps, mothers. It's also very hard to recognise (there are no control groups in the real workplace) and, therefore, contest. But as the next chapter shows, this newer, kinder form of discrimination hasn't *replaced* the old, intentional variety. These days, they can work together.

> Let the . . . women carry on their crusade for a generation or
> two more; let men meet women as competitors for 'economic
> independence' and in the hard fight of wringing a living from the
> world; let men meet women in the fierce struggle of political life;
> let the screeching rowdyism of the militant suffragettes go on
> and grow worse; but, above all, let the feminist programme of
> greater sex liberty for women, with its demolition of wifehood
> and the home, be carried through; then will women indeed find
> that the knightliness and chivalry of gentlemen have vanished,
> and in their stead will arise a rough male power that will place
> women where it chooses.
>
> —William T. Sedgwick, professor of biology and
> public health at MIT (1914)[1]

Unlike many of his contemporaries (who, as we will see, made pessimistic predictions such as voting-induced insanity or ovaries shrivelled from overeducation) Sedgwick was actually onto something. This threatening passage offers women a choice between the carrot and the stick, or what social psychologists Peter Glick and Susan Fiske refer to as benevolent and hostile sexism, respectively. So long as women stick to their traditional caring roles, they can bask in the stereotype of the 'wonderful' woman – caring, nurturing, supportive and the needful recipients of men's knightly chivalry – without whom no man is complete. But the woman who seeks nontraditional high-status and high-power roles risks triggering the hostile sexism that 'views women as adversaries in a power struggle'.[2] Hostile discrimination against women in the workplace is intentionally and consciously done. It can involve 'segregation, exclusion, demeaning comments, harassment, and attack.'[3] It's still with us.

Professor Sedgwick, it should be said, probably did not anticipate that such hostilities would still be being directed at women a century later. Not because this would seem to be time enough for everyone to get used to the idea of women asking for a share of the jobs that men had allocated to themselves. Rather, because he predicted that men would soon call a halt to the whole feminist endeavour 'and, putting the women back in their homes, say: "That is where you belong. Now stay there."'[4]

While we might think this kind of explicitly held attitude a relic of the past, legal scholar Michael Selmi argues that a 'lingering bias' towards precisely this point of view – that women are caregivers and men are breadwinners – can manifest itself in workplace discrimination. He suggests that 'our perceptions of discrimination may have changed more than its reality, and there is certainly strong reason to believe that intentional and overt discrimination remains a substantial barrier to workplace equality for women.'[5] He bases this conclusion on a review of class-action employment discrimination cases, especially in the securities and grocery industries, from the nineties to the early years of this century (the wheels of justice, as we all know, turn slowly). A common theme in these cases (all of which settled), Selmi argues, is the exclusion of women from higher-paying positions with greater promotional opportunities; and these discriminatory decisions were based on unexamined, stereotyped assumptions about female employees' work preferences. Women *prefer* those kinds of dead-end jobs because they fit better with their family commitments, the companies typically claimed in their defence when their happily fulfilled female employees filed lawsuits against them.

Yet as Selmi points out, the companies had no evidence that this was the case. Indeed, the aggressive, ambitious women working in the securities industry, in particular, 'should have provided an important counterweight to the underlying stereotypes'.[6] Those on the top rungs were not unconsciously seeing women as slightly less qualified for better roles. They were *consciously* deciding, without giving women a chance to decide for themselves, that these more generously remunerated (and, ironically, possibly more-flexible) jobs were for men. Several large retailers in other industries have been hit with similar allegations, Selmi notes.

Beyond gender stereotypes, homophily (a psychological tendency captured by the old adage that 'birds of a feather flock together') can often create barriers for minority workers. A recent interview study of current and former Wall Street professionals revealed that they took it for granted that client organisations made up primarily of white men will prefer to deal with other white men. This meant that women and nonwhite professionals were excluded from the most lucrative jobs in the securities industry and were instead 'concentrated in jobs without client contact and in client-contact jobs that generate less revenue.'[7] Social exclusion may also hold back women who work in other traditionally male domains. The Athena Factor report mentioned earlier found that women in corporate SET jobs were being denied the sort of insider information that they needed to get ahead. One Silicon Valley participant, a major player in the technology industry, gave herself a male alias and discovered that the emails that 'Finn' received were completely different from those sent to 'Josephine'. Finn got the scoops and Josephine got the 'pap'. The report authors also describe 'alpha male techies' as combining poor social skills with an arrogant sense of male superiority. 'One focus group participant described a recent uncomfortable experience. A male colleague walked up to a group where she was the only female. The man shook the hand of every man but avoided contact with her. "I could feel his anxiety in assessing how to handle greeting me," she noted. "But he also didn't think I was important. So in the end he just chose not to deal with me."'[8] This anecdote suggests a workplace environment that tolerates a deep disrespect for women. No intellectually functioning adult, however meanly endowed with social skills, can have failed to learn the social rule that it is rude to shake hands with every single person in a group except one. No less remarkably rude is the behaviour of a surgeon remembered by Kerin Fielding, one of Australia's few female orthopaedic surgeons. She recalls having had 'many battles' during her training, including one particular surgeon who refused to work with her. When Fielding met the same man years later he condescendingly enquired whether she had many patients, insultingly adding, 'It's just toes, fingers, I suppose.'[9]

Unfortunately, the problem for women of being excluded does not end when they leave the office. Depressingly, it is still the case that in many industries it gets worse. At first glance, a round of golf and a trip to the local lap-dancing club may seem to have little in common. They are both leisure activities, it's true, but one is conservative, traditional and may even entail the wearing of Argyle socks, while the other involves naked women rubbing their genitalia against the fly region of a man's pants. What they share, however, is an environment that provides ample scope for excluding women from valuable client networking opportunities.

In business-to-business sales, developing a good personal relationship with the client through out-of-office socialising is a vital part of the work. Unfortunately, two of the more popular venues for client entertaining – golf courses and strip clubs – both offer ample scope to keep women away from the networking action. Many golf courses are run around the principle that there would be something unnatural and absurd about women playing golf at the same time as men – or even at all. Even when women and men can play together, the different tee boxes used for the two sexes keep them somewhat separate. 'Many women reported that men used the different tee boxes to leave them behind on the course or to require them to ride in a different golf cart. . . . In essence, they used the different tees as a way to exclude women even when playing with them', report University of Michigan sociologists Laurie Morgan and Karin Martin, who studied the experiences of female sales professionals.[10]

Another popular entertainment venue that creates 'enormous challenges' for professional saleswomen, Morgan and Martin found, is the strip club. Perhaps unsurprisingly, male colleagues and clients are reluctant to have a woman from the office at such venues, spoiling their fun by reminding them that women are more than simply bodies to be looked at. The saleswomen 'described over and over again being told not to come, not being invited, and even being deceived as the men snuck out to a strip club.' But these women were determined. Even though being there was often extremely awkward for them ('they feel different, out of place, and embarrassed'), they went. They didn't want to miss out on the valuable opportunity to socialise with important clients.[11]

And then there are the lap-dancing clubs. A survey by the UK Fawcett Society, based on anonymous testimony from city workers, found that it is 'increasingly normal' for clients to be entertained at these kinds of venues.[12] Expected, even. Regarding the issuing of a licence to a lap-dancing club in Coventry, England, a 'leading businessman' argued to the council that '[i]f Coventry has aspirations to be a major business area, then it has to have a quality adult entertainment area, and that would include a lap dancing club.'[13] How on earth did men ever manage to get business done in the days before establishments where they can pay to have their penises massaged by the genitalia of a naked woman? 'The City guys are a very large part of my market', commented Peter Stringfellow, shortly after investment bank Morgan Stanley fired four U.S. employees for visiting a lap-dancing club while attending a work conference.[14] The Web site for his eponymous 'world famous nude dancing clubs' has a Web page specifically devoted to corporate events, which describes the Stringfellows clubs as 'perfect for your discreet corporate entertaining'. The copy excitedly asks, 'OK so you've just done the big deal, or you're about to do the deal but they need that extra little push. So tell me, where are you going to take them to clinch the deal???' By way of answer, it displays a picture of '[y]our perfect private party table'. The said table differs from conventional ones in that a pole rises up from its centre. No doubt any female investment banker attending the deal-clinching moment would be thrilled by the convenience of being able to prepurchase, with her company credit card, String-fellows Heavenly Money (depicting a nude woman clasping a pole) to tuck into the garter of the naked woman gyrating between the soup bowls.[15] How 'perfect' to be able to dine with her colleagues, network with important clients, and all while enjoying the view of another woman's genitals. Or perhaps she'll have a headache and stay home. Stringfellows is by no means unusual in accommodating the corporate market. The recent Corporate Sexism report by the Fawcett Society found that 41 percent of the UK's lap-dancing clubs specifically promote corporate entertainment on their Web sites, and 86 percent of the London clubs offer discreet receipts, which enable the cost of the evening's activities to be claimed as a company expense.[16]

It's not hard to see that – whatever your moral take on strip joints and lap-dancing clubs – using them as corporate entertainment serves to exclude women. Said one saleswoman working in the industrial sector, 'they will never have a woman work in that group because part of their entertainment is to take people to these topless bars.'[17] With perhaps as many as 80 percent of male city finance workers visiting strip clubs for work,[18] 'women in the world of business . . . are confronting a new glass ceiling created by their male colleagues' use of strip clubs', points out political scientist Sheila Jeffreys.[19] Or, as journalist Matthew Lynn put it:

> In effect, just as their fathers might have taken clients to one of the gentlemen's clubs of Pall Mall, so brokers today take their business associates to see lap dancers. The old gentlemen's clubs banned women – some still do – whereas the lap-dancing establishments merely intimidate them.[20]

And this brings us neatly to what is perhaps the most effective way to express hostility towards women in the workplace: sexual harassment. Michael Selmi also reviewed numerous sexual harassment class-actions (all but one of which settled), focusing on cases in the automotive and mining industries where women sought access to some of the best-paying jobs in the area. He describes 'an all too familiar litany of harassment – groping, grabbing, stalking, pressure for sex, use of sexual language and pornography, men exposing themselves and masturbating on women's clothes.' Nice. The sheer crudity of the behaviour suggests that these kinds of harassing behaviours stemmed not from the erotic charge of having women around, but rather provided a way of 'creating an environment that conveyed express hostility to women' and 'disciplining women who sought to infiltrate previously all-male workplaces.'[21]

Nor are the environments of male-dominated white-collar professions necessarily ones that make women feel that they are welcomed as professionals worthy of equal respect. The securities industry lawsuits often included allegations of 'pervasive sexual harassment' (as well as the allegations of mistreatment of women in promotion, training, mentoring, and the assignment of lucrative accounts). While Selmi acknowledges that it's tricky to draw conclusions from cases that have settled, which was the situation for all of the securities lawsuits he discusses, he argues that 'it is equally clear that the allegations all appear to have been substantiated at least to some significant degree.'[22]

The Athena Factor report found that 56 percent of women in corporate science jobs, and 69 percent of women in engineering, had experienced sexual harassment. 'Locker-room language and sexually explicit taunts are standard and hard to take.'[23] And almost all of the ninety-nine female medical residents at Southern University interviewed by sociologist Susan Hinze reported experiencing 'sexual harassment that makes the workplace intimidating, hostile, or offensive'.[24] Surgery, the most prestigious branch of medicine, offered by far the most hostile environment to women. Yet the recurring theme in Hinze's follow-up interviews with the residents was not anger, or even victimhood, but whether women were being overly sensitive to sexist and demeaning treatment. For example, a woman who was repeatedly patted on the behind by an anaesthesiology-attending physician wondered whether the discomfort this caused her was a sign she was being too sensitive. She deliberated whether, if she mentioned it, her colleagues would say, 'whooa, she's a real bitch, she's sure uptight, she's sure sensitive . . .' Another resident was furious when a male faculty member, seeing her shivering, said 'Oh, I wish I could just take you on my lap like I would my little girl, and hold you tight and warm you up.' As she angrily pointed out to the interviewer, 'I'm not here to remind him of his daughter. I've gotten this far in life and I remind him of his little daughter?' But other people reassured her that there was nothing objectionable about his comment. And female medical students offended by one surgeon's habit of referring to them as 'little girl' were denounced as 'hypersensitive' by a male peer who suggested that women's 'nerve endings' are 'absolutely naked' and thus primed to take offence.[25]

But contrary to this opinion, the female residents actually seemed to be working very hard to, as Hinze suggests, 'downplay the incidents and view them as a "normal" part of a bruising training experience' (which indeed it is for men and women alike), and to either ignore it ('I'm in surgery; I can't sweat the small stuff') or see the need for change in *themselves* rather than in those who harassed them. As one resident warned, 'if you blow up every little comment that somebody makes to you . . . you're too sensitive.' One surgery resident described the experience of discovering in the restrooms an explicit cartoon of herself, bent over, and her mentor engaged in sexual intercourse. Another resident had added an arrow and the comment that he wished he could be in the latter's position. The woman recalled to Hinze:

> I thought, this just really sums up . . . my position in the department of [name removed] surgery, something I've worked for for a lot of years, not my whole life, but a lot of years, and they reduce all my hard work and all my sacrifice and my brains and my technical abilities and everything that I've done to this,

you know, like this is how they perceive, you know, me. [R
becomes visibly upset, begins crying]

She filed no complaint but looked to herself to adapt to the
hostile environment ('I might as well just get over it') without
any expectation that she should not have to deal with this kind of
treatment at work ('that's how men are').[26]

This example underscores one benefit to women of ignor-
ing, shrugging off or refusing to identify hostile discrimination.
Frankly, it is not kind to the self-esteem of women to be reminded
by sexual harassment that 'they are not equal to men in the workplace, that they are, still, after all their gains, just women'.[27] But
also, of course, publicly naming discrimination of any kind is neither easy nor guaranteed to bring about positive change nor something anyone does lightly when career, reputation and (if lawyers
get involved) savings are at stake. Even responding to a single
instance of sexual harassment is harder than one might think.
Imagine if, at an interview for a research assistant job, the male
interviewer asked you (a woman) questions like *Do people find you
desirable?* and *Do you think it's important for women to wear bras to
work?* How would you respond? Would you refuse to answer? Get
up and leave? Report the interviewer? These are all actions far
easier to implement in theory than in practice. When women were
put in this extraordinary situation for real, not one of the twenty-
five women in the study responded in these ways. Mostly, they just
smiled politely, and answered the questions.[28]

Things have improved since Professor Sedgwick's prophecy. In
1869, the dean of the Woman's Medical College of Pennsylvania
proudly brought her students to the Saturday teaching clinics in
general surgery at the Pennsylvania Hospital. She had, for years,
been seeking permission for her female students to be able to attend
and benefit from observing the great clinicians at work. At last, the

managers had agreed. But the young women did not receive a hospitable welcome. As reported in the Philadelphia *Evening Bulletin*:

> The students of the male colleges, knowing that the ladies would be present, turned out several hundred strong, with the design of expressing their disapproval of the action of the managers of the hospital particularly, and of the admission of women to the medical profession generally.
>
> Ranging themselves in line, these gallant gentlemen assailed the young ladies, as they passed out, with insolent and offensive language, and then followed them into the street, where the whole gang, with the fluency of long practice, joined in insulting them. . . .
>
> During the last hour missiles of paper, tinfoil, tobacco-quids, etc., were thrown upon the ladies, while some of these men defiled the dresses of the ladies near them with tobacco juice.[29]

Needless to say, the working environment for women is far better now than it was a hundred years ago. Equal opportunity law obviates any need for special pleading for women to receive the same educational opportunities as men, and female professionals and workers are commonplace, rather than controversial. And yet, compared with having ones backside repeatedly fondled by a surgeon, feeling obliged to network clients at a strip club, or having one's clothes masturbated upon, a bit of tinfoil in the hair and tobacco juice on the dress seems almost gentlemanly by comparison. As Michael Selmi notes, the many examples of overt discrimination against women in the workplace might be dismissed as 'isolated incidents'. Yet he argues that it would be 'a mistake to dismiss . . . as aberrational in nature' these examples of 'overt acts of hostility and exclusion based on stereotypes regarding women's proper roles or abilities in the workplace.'[30] Of course, not all mistreatment or harassment is directed at women in traditionally male occupations, or at women, and not all women are harassed. (One expert estimates that perhaps 35 to 50 percent of women have been sexually harassed at some point in their working lives.)[31] But the hostilities, sexism and demeaning indignities faced by some women in the modern workplace suggest that old ideas about the appropriate sphere of women continue to linger in many minds – a theme that continues in the next chapter, when we return home from work.

> S. and I have decided to get married next year when we get
> through medicine . . . I told him I didn't know a thing about
> housekeeping, and he said why should I? That he could see
> no more reason for a woman's liking cooking and dishwashing
> than for a man's liking them. That since our education has been
> precisely similar . . . there would be no justice at all in my having
> to do all the 'dirty work'. . . . So we have decided that one week I
> shall take over all the duties connected with the running of our
> house and the next week he will . . . I was so happy I couldn't
> speak . . . We are going to divide up the care of the children
> exactly as we divide the housework.
>
> —Dr. Mabel Ulrich, Johns Hopkins graduate (1933)

This hopeful arrangement was declared a 'no go' after just a few months, as Regina Morantz-Sanchez reports in *Sympathy and Science*. 'We have given up the 50-50 housekeeping plan. We tried for a month, but by the end of one week I knew S. is a fearful mess as a housekeeper. . . . Could never remember the laundry. . . . But then of course he is busy and I am not.'[1]

Dr. Ulrich was, in the first half of the twentieth century, running up against the implacable psychological force of the middle-class marital contract. According to this traditional and highly familiar arrangement, the husband is the breadwinner and works outside the home to provide financial resources for the family. In return, his wife is responsible for both the emotional and household labour created by the family: keeping everyone happy, the house clean, meals cooked, clothes laundered, and children reared; either by her own hand or by proxy. Because *this* becomes the woman's job once married, employers were perfectly entitled to fire or refuse to employ married women – a situation that remained perfectly legal in the United States until 1964.

Both the breadwinner and the caregiver roles are, of course, necessary. Without the breadwinner there is no money for food. But without the caregiver, the food is not cooked; there is no clean plate on which to place it; and the children are living naked, filthy, and wild in the garden, communicating by way of a primitive system of grunts. The 'separate spheres' of men and women – his public, hers private – were seen as complementary and equal, but in an *Animal Farm*-ish some-spheres-are-more-equal-than-others sort of way. When I say 'head of the household', you immediately know to which spouse I refer (and it's not 'Mrs. John Smith'). That his was the final word was enshrined in law until surprisingly recently. Not until 1974 did US legislation require that married women be able to apply for credit in their own names. And it was only in 1994 that it became possible in the eyes of the law for a British husband to rape his wife. I mention these points not to lower the mood, but simply to highlight the asymmetry of power and status in the traditional marriage contract.

Contemporary women seem to be barely more successful than Mabel Ulrich in persuading their partners to step into the traditionally female private sphere. My husband and I can both enthusiastically attest to the difficulties inherent in attempting an egalitarian marriage – particularly when children are involved. You have heard, no doubt, the saying that the personal is the political. Based on his own experiences within a marriage in which we struggle against convention to split things equally, my husband has developed his own, expanded version of this motto. As he would state it, 'The school drop-off is the political, the staying home when the kids are sick is the political, the writing of the shopping list is the political, the buying of the birthday presents is the political, the arranging of the baby-sitter is the political, the packing of the lunch boxes is the political, the thinking about what to have for supper is the political, the remembering of the need to cut the children's toenails is the political, the asking of the location of the butter dish is the political . . .' You get the idea. Some day, I must ask him what it's like to be married to someone who, eyes narrowed in thought, peers at him over the tops of sociology articles with titles like *Who Gets the Best Deal from Marriage: Women or Men?* We've had our disagreements, of course. When, for example, are a few dirty cups a symbol of the exertion of male privilege, and when are they merely unwashed dishes? But however predisposed the research for this book has made me to see inequality where perhaps there is only a cluttered sink, my beleaguered husband can at least take comfort in knowing that, thanks to that very same research, I know just what a rare jewel he is.

In families with children in which both spouses work full-time, women do about twice as much child care and housework as men – the notorious 'second shift' described by sociologist Arlie Hochschild in her classic book of that name.[2] You might think that, even if this isn't quite fair, it's nonetheless rational. When one person earns more than the other then he (most likely) enjoys greater bargaining power at the trade union negotiations that, for some, become their marriage. Certainly, in line with this unromantic logic, as a woman's financial contribution approaches that of her husband's, her housework decreases. It doesn't actually become *equitable*, you understand. Just less unequal. But only up to the point at which her earnings equal his. After that – when she starts to earn more than him – something very curious starts to happen. The more she earns, the more housework she does.[3] In what sociologist Sampson Lee Blair has described as the 'sadly comic data' from his research, 'where she has a job and he doesn't . . . even then you find the wife doing the majority of the housework.'[4]

What on earth could be behind this extraordinary injustice in which she returns home from a hard day at work to run the vacuum cleaner under his well-rested legs? A few popular writers have made some creative suggestions. John Gray, author of the *Men Are from Mars, Women Are from Venus* books, has recently made a valiant stab at arguing that performing routine housework chores is actually selectively beneficial to women, including – if not especially – those with demanding jobs. His idea (which to my knowledge has not been empirically tested) is that because the modern working woman has removed herself from her traditional home sphere with its babies, children and friends on whom to call with a pot roast, she has dangerously low levels of oxytocin coursing through her blood. (Oxytocin is a mammalian hormone associated with social bonding and social interactions.) Thankfully, however, 'nurturing oxytocin-producing domestic routine duties like laundry, shopping, cooking, and cleaning' are available in plentiful supply. Phew! Such chores, however, have a very ill effect on men. For them, the priority is 'testosterone-producing' tasks – for without the stimulating rush of that sex hormone, men become little better than limp rags (and not even ones that then wipe themselves along the countertops). Thus, 'putting things back together after a flood or disaster' is testosterone-producing, but '[t]o expect him to join in and share each day in her daily routines as a helper would eventually exhaust him.' It's hard not to be a little cynical when Gray argues that it is in deference to his male neuroendocrinological status that when he helps with the dishes it should fall to 'others [to] bring plates over, put things away, and clean tabletops'. As he explains, '[h]aving to ask your partner each time whether this food should be kept, and remembering where she wants things to be put away, can be a bit exhausting for a man'.[5] One can only hope that Mrs. Gray finds it gratifyingly oxytocin producing to have to remind her husband where the plates are kept.

Or, there is the neuroscientific explanation offered by 'social philosopher' Michael Gurian in his popular book *What Could He Be Thinking?* In the chapter entitled 'The Male Brain at Home' we learn that because '[t]he female brain takes in more sensory data', a woman is more likely to 'neurally register the bit of paper, the dog hair, the children's toy shoved into the couch'. The 'female brain' is also 'more likely to sense the book that is awry on the coffee table, the dust on the end table, the bed not made as she'd like it'.[6]

If you are somehow sceptical of the notion that high-earning women do more housework because of an internal drive to maintain the highest possible oxytocin levels, while unemployed husbands carefully protect their own physiological state by giving the laundry pile a wide berth, or are simply neurally less capable of sensing it, then sociologists have an alternative explanation that you may find more satisfying. They refer to this curious phenomenon as 'gender deviance neutralisation'.[7] Spouses work together to counteract the discomfort created when a woman breaks the traditional marital contract by taking on the primary breadwinning role. A fascinating interview study conducted by sociologist Veronica Tichenor revealed the psychological work that both husbands and their higher-earning wives perform to continue to 'do gender' more conventionally within their marriage, despite their unconventional situations.[8] For example, as predicted by the quantitative surveys, most of the higher-earning wives also reported doing the 'vast majority' of both domestic labour and childrearing. Sometimes this was resented and a point of contention. But others seemed to 'embrace domestic labour as a way of presenting themselves as good wives.' As Tichenor points out, what this means is that 'cultural expectations of what it means to be a good wife shape the domestic negotiations of unconventional earners and produce arrangements that privilege husbands and further burden wives.'

Tichenor also surmised that in decision making the women were deferring to their husbands in 'very self-conscious ways' because they didn't want to be seen as powerful, dominating, or emasculating. The couples also redefined the meaning of 'provider' so that the men could still fall within the definition. While in the conventional couples the provider was the person who brought home the biggest paycheque, among the other couples the men's management of the family finances, and other noneconomic contributions, were considered part of providing. Thus it was that Bonnie, earning $114,000 a year and married to a man earning $3000, could nonetheless argue that they were 'both providers'. Interestingly, these women were often very aware that their greater income didn't bring them the same power within the relationship as it would a man in a more conventional marriage.[9]

These psychological scrambles reveal the strength of the push to maintain gendered roles, Victorian-style, within marriage. As Michael Selmi has pointed out, even though more than 80 percent of people born between 1965 and 1981 support the idea of equal caregiving, actual progress towards this goal has been 'glacial'.[10] Why is it still so hard, and so rare? Mabel Ulrich had a suggestion:

> A man, it seems, may be intellectually in complete sympathy with a woman's aims. But only about ten per cent of him is his intellect – the other ninety is emotions. And S.'s emotional pattern was set by his mother when he was a baby. It can't be so easy being the husband of a 'modern' woman. She is everything his mother wasn't – and nothing she was.[11]

Dr. Ulrich's suggestion dovetails beautifully with the curious split often seen between the gender-equal values people consciously endorse and the automatic gender associations that, through their influence on thought and act, can undermine those beliefs.[12] For example, one study found that a group of childless female college students reported that they valued a college education more than motherhood. Yet on the IAT, they found it easier to link self words (like *I*, *me*, and *self*) with pictures of the paraphernalia of motherhood (such as cribs and strollers) than with images of college (like graduation gowns and binders).[13] These automatic attitudes have an impact on our behaviour, over and above that of the values we consciously report.[14] One study even found that only these were correlated with women's career goals. Laurie Rudman and Jessica Heppen measured how strongly a sample of young women implicitly linked romantic partners with the sort of shining knight heroism of fairy tales, and also asked them directly what they thought of such sugar-coated fantasies. Remarkably, it was the strength of a woman's implicit romantic fantasy associations, rather than any no-nonsense views that she personally endorsed, that correlated (negatively) with her level of interest in achieving high-status and educationally demanding occupations.[15] Research into the development of automatic associations is still in its early stages, but preliminary findings suggest that, just as Ulrich proposed, they may be most strongly impacted by early childhood experiences.[16] In which case, as we'll see in the third part of this book, it is hardly surprising that implicit gender associations are so traditional.

People can and do act against the implicit mind and more in line with their consciously endorsed values. But if *her* implicit mind, or her social identity as a mother or wife, triggers her to load the washing machine, unload the dishwasher and put away the children's clothes – while *his* implicit mind is not so helpful on such matters – then before you know it you are engaged in what sociologists describe as 'actively negotiating and continually challenging prevailing gendered assumptions about work and family roles' and the rest of us call 'plain old arguing'.[17]

Or perhaps it is not even as subtle as this. Powerful social norms still regard home and children as primarily *her* responsibility, even if he is now expected to help. A marvellous poster, put out by the National League for Opposing Woman Suffrage in the UK, depicts a husband returning to 'a suffragette's home'. The room is in cheerless disarray, the weeping children have holes in their socks, and a fuel-less lamp emits not light, but smoke. The only evidence of the errant wife and mother is a 'votes for women' poster on the wall, on which is pinned a note bearing the callous words, 'back in an hour or so'. Just substitute the words 'working mother' for 'suffragette' and the poster could still be used today to great effect. While there are entire chapters – books, even – devoted to the issues of being a working mother, rare indeed is it to come across even a paragraph in a child-rearing manual that addresses the conflicts of time and responsibility that arise from being a working father.

This social norm puts women in a weak negotiating position. Anecdotally, many mothers I have spoken to have already eliminated from their mental decision-space – as if they simply did not exist – any work choices that would require their husband to take more (or even any) responsibility for the children. Needless to say, this immediately sweeps a number of options off the table. Sometimes there might be genuine practical or financial reasons for this. However, the head begins to swim when you start to look into the circularity behind such impasses.[18] One legacy of the neat breadwinner/caregiver division of labour is an expectation of the 'zero drag' worker who, because home and children are taken care of by someone else, can commit himself fully to his job. This expectation will not change, so long as women continue to cover family responsibilities. Of course, some jobs really aren't flexible. But it *is* curious just how bendy and stretchy a woman can make a job that appears a good deal more rigid and inflexible when pursued by a male. *Halving It All* author Francine Deutsch describes two couples she encountered. In one couple, he was a college professor and she was a physician, and in the other couple she was the college professor and he the physician. But in both cases, 'both the husband and wife claimed the man's job was less flexible.'[19] Then, there's the motherhood penalty (in addition to other gender-based pay inequalities) that increases the financial clout of his salary relative to hers.[20] Finally, the more a woman adapts her career to family commitments, and the longer the accommodation goes on, the wider the gap between his and her salary and career potential becomes. And so it becomes increasingly rational to sacrifice her career to his.

We begin to see how any hazy notions of an equal partnership that couples might once have held begin to seem like nothing but youthful folly.[21] Mabel Ulrich spent several years trying to juggle a private medical practice (which she eventually gave up), family, and children. Having turned down a job offer to save her husband the inconvenience of having to move his medical practice, she wrote, 'I don't believe a woman's work is ever so important to her as a man's is to him.'[22] Was this merely a psychological Band-Aid that Ulrich applied to the wound of her disappointingly unequal marriage? Or, as proponents of hardwired sex differences would suggest, had her abstract feminist ideals been dislodged by biological reality? Louann Brizendine, for example, suggests that the female brain responds to breadwinning versus family conflict 'with increased stress, increased anxiety, and reduced brainpower for the mother's work and her children', and that combining motherhood with career gives rise to a neurological 'tug-of-war because of overloaded brain circuits.'[23]

Overloaded brain circuits . . . or overloaded to-do list? Brizendine's claim that 'understanding our innate biology empowers us to better plan our future'[24] is not one I found especially compelling. I suspect most working mothers find other things more helpful: such as workplaces that are family friendly, and fathers who do the kindergarten pick-ups, pack the lunch boxes, stay home when the kids are sick, get up in the night when the baby wakes up, cook dinner, help with homework and call the paediatrician on their lunch hour. In fact, these turn out to be important absences in the lives of the so-called new traditionalist women who opt out of their often prestigious, lucrative and hard-earned careers to devote themselves to home and family. Their choice is usually attributed to the pull of women's different internal drivers. And yet sociologist Pamela Stone's detailed interview study of fifty-four such women, reported in her book *Opting Out? Why Women Really Quit Careers and Head Home*, reveals a fascinating and complex picture, and one in which gender inequality at home (alongside all-or-nothing workplaces) was a major factor in most interviewees' decisions to trade in the very successful careers that they loved. Their husbands, who also had demanding careers, were often described by their wives as being 'supportive' and giving their wives a 'choice'. But none provided a *real* choice to their wives by offering to adapt their own careers to family demands:

Women and their husbands appeared to perceive the latter's responsibility as limited to providing the monetary support to make it possible for their wives to quit, *not* to helping wives shoulder family obligations that would facilitate the continuation of their careers. 'It's your choice' was code for 'It's your problem'. . . . Veiled behind the seemingly egalitarian rhetoric of 'support' and 'choice,' husbands were in effect giving their wives permission to quit their careers, and signaling at the same time that women's careers were not worthwhile enough to merit any behavioral changes on their (the husbands') part.[25]

And although we tend to think that, perhaps because of hormones, there is something natural about fathers being more hands-off, biology offers us a lot more flexibility than we might think. Hormones are not simply internal drivers that pull us towards particular sorts of environments and behaviour: the influence works in the other direction, too. Stimuli in the environment – whether it is a baby, a success at work or a touching and moving segment on *Oprah* – can trigger hormonal changes.[26] Our hormones respond to the life we lead, breaking down the false division between internal biology and our external environment. And so, it should be little surprise to learn that it is not just mothers' hormones that change during the transition to parenthood, but fathers', too. (Although there is rather little research in this area, testosterone levels, for example, seem to be suppressed around the time of birth, while prolactin – which as the name suggests is a hormone implicated in lactation – increases.)[27] In her study of equal sharers – that is, mothers and fathers who equally share the responsibilities and pleasures of homelife – Francine Deutsch found that equally sharing fathers had developed the kind of closeness to their children we normally associate with mothers. Said one father of teenage girls, who 'expressed what many [equally sharing fathers] felt: "A lot of things I would change in my life. (Parenting) I wouldn't consider changing. It's the best thing I've done in my life."'[28]

And if this fails to convince, consider the rat. Male rats don't experience the hormonal changes that trigger maternal behaviour in female rats. They *never* normally participate in infant care. Yet put a baby rat in a cage with a male adult and after a few days he will be caring for the baby almost as if he were its mother. He'll pick it up, nestle it close to him as a nursing female would, keep the baby rat clean and comforted and even build a comfy nest for it.[29] The parenting circuits are there in the male brain, even in a species in which paternal care doesn't normally exist.[30] If a male *rat*, without even the aid of a William Sears baby-care manual, can be inspired to parent then I would suggest that the prospects for human fathers are pretty good.

Contrary to the idea of shared care as a modern, misguided fad, contemporary fathers may be less involved with their children than they were two to three hundred years ago. From the few available historical scraps of information about fatherhood in early America, historian John Demos has suggested 'a picture, above all, of active, encompassing fatherhood, woven into the whole fabric of domestic and productive life. . . . Fathering was thus an extension, if not a part, of much routine activity'.[31] When in the nineteenth century men's work increasingly moved outside the home, stories of the time 'picked up the tension' between career and home life. Demos describes a fictional father from the 1842 edition of *Parents' Magazine* (even the name is more progressive than the majority of titles today) who is so busy that he can no longer get home in time to conduct the family prayers. In the end, the father is brought 'back to his senses and his duty: "Better to lose a few shillings," he concludes, "than to become the deliberate murderer of my family, and the instrument of ruin to my soul."'[32] The question of the most appropriate care of the soul is well beyond the scope of this book. But hardwired accounts of gender that regard almost all men as single-mindedly career focused ignore gathering signs that some men no longer wish to be that instrument of ruin and would enjoy more time for family, friends and community.[33]

Whether this would rescue their souls, I cannot say. But one thing is certain. It would better enable these men, in stark contrast with Mabel Ulrich's husband, to do the laundry. And laundry *is* important. As Gloria Steinem recently reminded a journalist, 'The idea of having it all never meant doing it all. Men are parents, too, and actually women will never be equal outside the home until men are equal inside the home.'[34]

I s it time to crack open the champagne in celebration of the successful completion of Gender Equality 2.0, a revised version of equality in which men and women are not equal, but equally free to express their essentially different natures? Western women have contraception, equal opportunity laws and the economic freedom to pursue fulfilment rather than the dollar. And yet women's and men's choices and paths in life still diverge. 'But', asks *Sexual Paradox* author Susan Pinker, 'is this a problem that should be fixed?'[1] Is it time to stop assuming that women and men should live similar lives?

I do have sympathy for this concern. Sometimes, just for fun, my building contractor husband and I briefly imagine what it would be like if we were forced to swap jobs. My husband, who can take up to an hour to compose an email message that reads like a missive from a ten-year-old French pen pal (*Dear Michael. How are you? Today it was very hot.*), visibly blanches at the idea of writing a book. And were my husband to suffer a fatal accident at the beginning of a renovation project that I would then have to complete, he would most likely expend his dying breaths in the ambulance dictating a memo along the lines of: *Cordelia: Don't forget, sewerage and electrical wiring* before *walls go up! I love . . . [gurgle, clunk]*. Society would not be a better and happier place were more people like my husband to write books, and more people like me to renovate houses. Perhaps women are simply intrinsically less able at, or less interested in, the male-dominated fields of science, technology, engineering and maths because these occupations are less suitable and rewarding for a brain that inclines towards empathising. And if the majority of women are wired to nurture civilisation rather than advance it, then it should be no surprise that relatively few take on the demands of the most prestigious and greedy careers, and rise to the top. If male and female nature pushes men and women, on average, towards both horizontal segregation (the clustering of sexes in different occupational fields) and vertical segregation (the greater number of men at the top levels of all occupational fields), then there does seem something rather pointless and counterproductive, I agree, about a target of perfect equality.

However, we should not throw up our hands in defeat too quickly. Gender Equality 2.0 justifies a status quo in which politics, wealth, science, technology and artistic achievement continue to lie primarily in the hands of (white) men. This is not by any means to denigrate the importance and value of the work women traditionally do, or feminine qualities of character. But it's worth considering philosopher Neil Levy's argument that the idea that women are predominantly hardwired for empathising while men are hardwired for systemising 'is no basis for equality. It is not an accident that there is no Nobel Prize for making people feel included.'[2] When a child clings on to a highly desirable toy and claims that his companion 'doesn't want to play with it', I have found that it is wise to be suspicious. The same scepticism can be usefully applied here.

In a *New Yorker* cartoon that for many years enjoyed pride of place in my office, a rat in a business suit is at his desk, talking on the phone. On the wall behind him is a lever and a light. With his feet perched comfortably upon his desk, the rat-businessman is saying, 'Oh, not bad. The light comes on, I press the bar, they write me a cheque. How about you?'[3] The basic psychological principle that people find it rewarding to be rewarded – whether it be through sincere praise, status, money, a new opportunity, a promotion, a round of applause or a really nice review in a newspaper – should not be forgotten. Everyone, after all, knows the thrill of pride that accompanies acknowledgement of a talent or a job well done. As children we demand it. (*Look at me, Mummy. Look . . . at . . . ME!*). And as adults, although we're rather more discreet about our need for appreciation, we nonetheless lap it up wherever it's available. (I don't *think* it's just me.) On coaching mornings at my local tennis club, everyone edges towards Simon, a coach of such endless invention and generosity that he can think of something genuinely enthusiastic to say (*But nice footwork, Cordelia*) even as the ball sails over the fence into the windshield of a passing car.

The general idea that 'people's preferences are not created *ex nihilo*: they are formed by the society they live in'[4] is an important one to apply to our thinking about the reasons behind continuing vertical segregation, for example. Despite the great gains of the past century, men's and women's experiences at work and home are not the same, for reasons that often stem from either unconscious or intentional discrimination. If we rewarded one group of rats with bigger and better food pellets as they pulled a well-oiled lever in the spacious and enviable corner Skinner box, would we think them more intrinsically interested in lever-pulling than a less privileged, perhaps even harassed, group of rats? The managers who don't get the promotions or salaries they deserve, the saleswomen and investment bankers who determinedly network at topless bars and lap-dancing clubs, and the corporate scientists who endure locker-room culture deserve proper acknowledgement of barriers that still have not fallen.

And this includes barriers at home. Women with children who decide *not* to adapt their careers to family life can look forward to paying a gender deviance tax that takes the form of extra housework, extra child care, and a psychological pussyfooting around his ego. Who knows what goes on in any individual relationship. Of course, there are exceptions. But the data from a study of faculty at the University of California are telling.[5] Female faculty with children report working fifty-one hours a week at their jobs and another fifty-one hours a week doing housework and child care – truly the second shift. That's a 102-hour work-week, accounting for more than fourteen hours per day. Add to this eight hours per day for sleeping, an hour for eating and basic hygiene, and by my calculations that leaves these women the grand total of twenty-six minutes a day for themselves. Faculty fathers, by contrast, put in only thirty-two unpaid work hours a week. This substantially lighter load not only enables them to put in an extra five hours a week at work, but to also enjoy a spare *two hours a day* to spend doing – well, who knows – while faculty mothers continue to launder, cook, test spelling, wash grubby faces and read bedtime stories. Behind every great academic man there is a woman, but behind every great academic woman is an unpeeled potato and a child who needs some attention. And women who climb the academic ladder don't just forfeit their leisure. They are much less likely to be married with children than male faculty (41 versus 69 percent, respectively) and, poignantly, twice as likely once in their postreproductive years to say that they would have liked more children. Put simply, the same career entails greater sacrifices for her than for him. So when a female academic who would like to have more than a few minutes for herself every day, as well as a family, jumps off the academic ladder and into a more flexible but dead-end second-tier research position, is it because she's intrinsically less interested in a demanding academic career or because there are only twenty-four hours in a day?

Likewise, our societies also offer a surprisingly poor test of the naturalness of horizontal segregation. Picture, if you can, a society in which men expect to find happiness not from work but from their family and friends. Imagine a place in which equal numbers of women and men, sitting attentively in the lecture halls of the computer science department, set themselves up for a financially secure future. This society is no feminist fantasy of the future. It is the Republic of Armenia. In the 1980s and '90s, the percentage of women in the largest computer science department in the country did not fall below 75 percent. Today, thanks to its increasing popularity among men (rather than declining popularity among women), Armenian women still make up close to half of computer science majors (and, anecdotally, their numbers appear to be high in many other former Soviet Republics[6]) – compared to about 15 percent in America. Hasmik Gharibyan, a professor of computer science at California Polytechnic State University, attributes the disparity to important cultural differences between the two countries. In Armenia '[t]here is no cultural emphasis on having a job that one loves'. In every one of her interviews, the young Armenians 'emphasized that the source of happiness for Armenians undoubtedly is their family and friendships, rather than their work'. Instead, for women and men alike, 'there is a determination to have a profession that will guarantee a good living and financial stability.'[7]

The strong representation of Armenian women in computer science is just one example of what is a rather surprising general pattern: there is *more*, not less, gender segregation of occupational interests in rich, advanced industrial societies than in developing or transitional ones. For example, a recent survey of forty-four countries found that as economic prosperity increases within developing and transitional countries, women are increasingly likely to turn away from degrees in engineering, maths and natural science (that lead to potentially more lucrative careers) and instead choose more feminine degrees in the humanities, social sciences and health. But in prosperous countries it is not economic prosperity that tracks sex segregation in degree choices, but differences in adolescent boys' and girls' attitudes towards maths and science. In richer countries, the greater the difference between boys' and girls' interest in science and maths, the greater the sex segregation.[8] Maria Charles and Karen Bradley, the survey authors, argue that a combination of an adequate baseline of material security (for most), together with a Western cultural emphasis on individual choice and self-expression, means that self-realisation in education is a culturally legitimate goal. This is especially true for people who might reasonably anticipate that their partner will take on the primary breadwinning role – namely, heterosexual women. (In fact it is interesting that, in the absence of the luxury of a male breadwinner, the occupational decision making of lesbians looks very similar to that of heterosexual men.)[9]

Susan Pinker interprets the occupational sex segregation in countries like the United States, Australia and Sweden as reflecting women's true preferences, unforced by financial concerns, family pressure or even governmental control. But as we've seen, occupational interests cannot be safely carried around inside the head, impervious to outside influence. We've *seen* the cultural cues that can so readily alter young people's interest in maths, science and other masculine pursuits. As Charles and Bradley argue, once males and females no longer have to chase the dollar as a top priority, they can 'seek to realize and express their true "selves"'[10] – but as you, I, and Charles and Bradley are aware, the boundary between the desires of that self and the gender beliefs and structure of the culture in which it develops and functions is permeable. Contrary to what you might expect, people from more gender-egalitarian countries are often *less* egalitarian when it comes to the gender stereotypes they typically endorse.[11] Charles and Bradley suggest that we in the developed West are 'indulging our gendered selves', and we've seen here a glimpse of how those selves become gendered. Cultural realities and beliefs about females and males – represented in existing inequalities; in commercials; in conversations; in the minds, expectations or behaviour of others; or primed in our own minds by the environment – alter our self-perception, interests and behaviour. These laboratory experiments are designed to simulate, in a controlled and tidy way, the far messier influences taking place in the real world. A sociocultural environment is not some cunningly contrived thing that only exists in social psychology labs. Don't look now, but you're in one right this moment.

Several researchers have suggested that the continual drip, drip, drip of gender stereotypes will, over time, really add up. For example, having observed the feminising effect of gender priming on women's interests, Steele and Ambady wonder whether 'our culture creates a situation of repeated priming of stereotypes and their related identities, which eventually help to define a person's long-term attitude towards specific domains.'[12] Likewise, sociologists Cecilia Ridgeway and Shelley Correll argue:

> [C]ultural beliefs about gender act like a weight on the scale that modestly but systematically differentiates the behavior and evaluations of otherwise similar men and women. While the biasing impact of gender beliefs on the outcomes of men and women in any one situation may be small, individual lives are lived through multiple, repeating, social relational contexts. . . . The small biasing effects accumulate over careers and lifetimes to result in substantially different behavioral paths and social outcomes for men and women who are otherwise similar in social background.[13]

These gendered paths and outcomes then become part of the social world that entangles minds – gendering the very sense of self, social perception, and behaviour that will then seamlessly become once again part of the gendered social world.

But it happens imperceptibly. And so we look for answers elsewhere.

PART 2

Neurosexism

> For two millennia, 'impartial experts' have given us such tren-
> chant insights as the fact that women lack sufficient heat to boil
> the blood and purify the soul, that their heads are too small,
> their wombs too big, their hormones too debilitating, that they
> think with their hearts or the wrong side of the brain. The list is
> never-ending.
>
> —Beth B. Hess, sociologist (1990)[1]

Twenty years later, and it's business-as-usual for that list. And somewhere near the top of it is 'too little foetal testosterone'. Or is it that males have too much of the stuff? At first, it might seem as though the tables have at last turned and that it's *males'* inherent deficiencies that are now under scrutiny. According to Louann Brizendine, for instance, the effect of male levels of testosterone on the foetal neural circuits is like nothing so much as the ravaging of a village by enemy soldiers:

> A huge testosterone surge beginning in the eighth week will
> turn this unisex brain male by killing off some cells in the
> communication centers and growing more cells in the sex and
> aggression centers. If the testosterone surge doesn't happen,
> the female brain continues to grow unperturbed. The fetal
> girl's brain cells sprout more connections in the communica-
> tion centers and areas that process emotion.

A consequence of this 'fetal fork', Brizendine explains, is that '[g]irls do not experience the testosterone surge in utero that shrinks the centers for communication, observation, and processing of emotion, so their potential to develop skills in these areas are [*sic*] better at birth than boys".[2] Girls, it seems – at least for the time being until we take a closer look at the data[3] – have not so much a deficiency of foetal testosterone as a lucky escape.

But really, this kind of portrayal is just new 'advertising copy' for the old stereotype of females as submissive, emotional, oversensitive gossips.[4] And a different, nicer way of saying that females' brains are designed for feminine skills rather than those necessary for excellence in masculine pursuits. Simon Baron-Cohen, willingly assisted by those who also popularise his work, has been doing a brilliant marketing campaign for foetal testosterone. It is rapidly becoming the must-have accessory for the budding hard scientist or mathematician. For example, in a recent article for BBC News, Baron-Cohen asks 'why, in over 100 years of the existence of the Fields Medal, maths' [equivalent of the] Nobel Prize, have none of the winners ever been a woman?' Over the course of the article, he circles around an answer . . . because women don't have the same testosterone-saturated in utero environment. So confident is Baron-Cohen about this link between foetal testosterone and mathematical ability that he expresses concern that a future, hypothetical prenatal treatment for autism that blocks the action of foetal testosterone might reduce 'that baby's future ability to attend to details, and to understand systematic information like maths'.[5]

This foetal testosterone certainly seems to be potent, sex-segregating stuff. So let's take a closer look, if we dare, at what it actually does.

At the beginning of life in the womb, male and female foetuses both have the same unisex primordial gonads.[6] But at around the sixth week of gestation, a gene on the male Y chromosome causes the male's primordial gonads to become testes. In the female the transformation is to ovaries instead. Shortly after, at about week eight of gestation, the testes of the male foetus start to produce large amounts of testosterone, often referred to as gonadal testosterone, which peaks at about the sixteenth week of pregnancy. (Researchers sometimes, more accurately, use the term 'androgens' rather than 'testosterone', because testosterone is one of several very similar hormones secreted from the testes, ovaries and adrenal glands, known as androgens.) By around the twenty-sixth week of gestation, there is once again little difference in testosterone levels between the sexes until another, smaller, testosterone surge in newborn boys that lasts for about three months. No one seems to be sure what this second, postbirth surge does. But the testosterone surge in utero is essential for bringing about male genitalia.[7] A genetic male without sufficient testosterone during this critical period will end up with feminised external genitalia, while genetic females with abnormally high testosterone in the same period are born with external genitalia that are masculinised – sometimes even to the extent that the baby girl is mistaken for a boy.

Such discoveries led to a brilliantly elegant idea known as the organizational-activational hypothesis. What if the same hormone involved in building male genitalia, a gift to be enjoyed for a lifetime, also permanently 'organises' the brain in a masculine way? (The other, activational, part of the hypothesis proposes that after puberty the circulating sex hormones activate these circuits.) Certainly, testosterone receptors have been found in many regions of the brain, in both males and females, and research with experimental animals is exploring how testosterone acts on the brain to influence its development.[8] And so, neuroendocrinologists have investigated the intriguing idea that prenatal testosterone organises the brain. They manipulate the hormonal environments of experimental animals during the critical period that brain organisation is thought to take place, and see what happens to their brains and behaviour.[9]

Probably the neatest support for the organizational hypothesis comes from songbirds like the zebra finch and canary, in which often the male sings but the female doesn't. In these species, the vocal control areas of the brain are much bigger and better in males, which makes perfect sense. What's more, treating female zebra finches to a male hormonal environment masculinises both their brains (in the vocal control areas) and their behaviour (they sing). Hormone, brain, behaviour – *snap!* (Actually, even here the picture can get a bit messy.)[10] But, while to perch on a branch and warble a song may be the best possible way to set yourself apart from the fairer sex if you happen to be a zebra finch, the same does not apply to the human case. And so this kind of result, fascinating though it is, can only get us so far.

When it comes to rat research, there are a few more points of contact. In rats, by the way, the surge of testosterone that appears to be involved in brain masculinisation actually takes place shortly *after* birth. Researchers have found that male rats castrated at birth are more similar to females in various ways, such as their propensity for aggression and how easily they become dazed and confused in a maze. Immediately, the cogs start to spin. Could prenatal testosterone in humans create permanent sex differences in the brain that lie behind gender differences in cognition and behaviour?

It's plausible but, as some researchers have pointed out, there are dangers in extrapolating from rats and birds to humans. Working from an implicit we're-all-God's-creatures framework that we do not apply when it comes to the right to not be killed and eaten, enjoy access to education or drive a car, there's a tendency (especially among some popular writers) to assume that what goes for the rat can be readily applied to humans.[11] Often, of course, this is the case. But while there *are* important similarities between all mammals great and small, there are also critical differences. As Melissa Hines points out (although she puts it rather less crudely), a penis is a penis, whether tucked between the legs of a rat or a man. Suitably scaled for size, it serves much the same function in both species, and the mechanism by which it's produced may be much the same in the two species. But a rodent brain, even expanded to suitably grand proportions, would serve a human extremely poorly indeed. Whereas in the human brain the so-called association cortices, devoted to complex and clever higher-order thinking, have taken over much of the available space, in the rat brain the association cortex has to squeeze in where it can among the neurons devoted to smell, sight, sound, touch and movement. It's for this reason that Hines cautions that 'one cannot assume that early hormonal influences on neural development in other mammals, particularly those involving the cerebral cortex, are preserved in humans.'[12] Likewise, the very point of the slur 'birdbrain' is to indicate that the thinking skills of the person in receipt of the insult are, in some important way worth commenting on, inadequate.

There are several other important dissimilarities, too, between how early hormones affect rats and humans.[13] All in all, some researchers think that rat data may not be very helpful in illuminating what goes on in humans.[14] That's not to say that the same principle doesn't apply – that foetal testosterone has some important effect on the brain. But it's wise not to extrapolate too enthusiastically from rats. So what about primates? Unlike rats, female rhesus monkey infants treated prenatally with testosterone are no more aggressive than untreated females. In fact, even normal female infants are no less aggressive than males when they are reared in a normal social group.[15] However, female infants experimentally treated prenatally with testosterone are keener than untreated females on rough-and-tumble play.[16] And when prenatal testosterone is blocked in males, early in gestation, these males are a bit less interested in rough-and-tumble play.[17]

Researchers hypothesise that the changes they see in behaviour as a result of their hormonal manipulations are brought about by testosterone-induced changes in the foetal brain (or, in the case of the rat, the neonatal brain). But I say *hypothesise* because it has proved harder than you might think, even in the relatively humble rat, to connect the dots between prenatal hormones, brain changes and behavioural change. For example, more than twenty-five years ago it was discovered that a certain region of the rat brain (part of the preoptic nucleus) is much larger in male rats than in female rats. Treating female rats with androgens early in life makes this region bigger, and depriving male rats of androgens prevents the normal male supersize appearance of the preoptic nucleus.[18] So far – hormone to brain – so good. But getting from brain to behaviour has proved a challenge. In 1995, the pioneer in this research, Roger Gorski, lamented, 'We've been studying this nucleus for 15 years, and we still don't know what it does.'[19] Nearly a decade later, neuroendocrinologist Geert De Vries pointed out again that scientists have 'not gotten an inch closer' to working out how this sex difference in the brain translates into behaviour. And not for want of trying.[20] Demand a story that includes a clear hormonal beginning, a neat neural middle, and a convincing behavioural end and the best that researchers have to offer involves a small area of the brain stem that innervates the penis. Without wishing in any way to denigrate the painstaking work of neuroendocrinologists (or, for that matter, the glory of the male machinery), so far they are falling way behind in the schedule of scientific discovery that Brizendine and others blithely attribute to them.[21]

And even here in the brain stem the story turns out to be much more complex than it first seems.[22] Celia Moore is a developmental psychobiologist at the University of Massachusetts who has put a lot of effort into trying to understand *how* early hormones bring about sex-typical behaviour in postnatal life. Is it really by way of some direct enduring effect on the brain, or is it possible that 'early hormones set all manner of processes into motion that could converge on behavioral differences days, weeks, months, or years down the road. What about those canines developing in young male rhesus monkeys? What about size differences resulting from early hormones? What about the genitalia? or odours, or other socially important cues?'[23]

Moore set out to investigate this very idea in the rat. Rat mothers lick the anus and genitals of their newly born pups, and Moore noticed that male pups are licked more than females. The reason for this, Moore discovered, is that mothers are attracted by the higher levels of testosterone in the urine of male pups. When Moore blocked the mothers' noses, they licked male and female pups equally; and female pups injected with testosterone were licked as often as their brothers. But most remarkable of all was the effect of this anogenital licking on the young rats' brains. When Moore stimulated the anogenital region of untampered-with female rats, using a paintbrush, the penis innervating nucleus in the brain stem got bigger (although not as big as the nucleus of a male rat). In other words, the sex difference in the nucleus size was not *just* due to neonatal testosterone, but was also influenced by the different maternal treatment of male and female pups.[24] Even our simple hormone-to-brain-stem storyline has a social subplot.

This should make us concerned that social experiences might also be involved somewhere along the path between hormones and behaviour, and this flags the danger of leapfrogging directly from one to the other. As Moore puts it, this approach leaves 'lots of unexplored territory and many possible pathways, perhaps convoluted ones, from the early hormones and end points of interest.'[25] We should bear this in mind when, in the next chapter, we look at this kind of research with humans (and other primates). Moore's work gives us a glimpse into the 'amazingly complex interaction of brain, hormones, and environment in creating behaviour. And if the process is complicated in rats, imagine how much more so it is in humans', as Rosalind Barnett and Caryl Rivers point out in their book *Same Difference*.[26]

But scientists are stout of heart. In the 1980s, Norman Geschwind and his colleagues suggested a very complex theory, part of which involved the idea that the high level of foetal testosterone experienced by males slows the growth of the brain's left hemisphere.[27] Geschwind went on to suggest that this leaves males with a greater potential for 'superior right hemisphere talents, such as artistic, musical, or mathematical talent.'[28] The Geschwind theory is the Teflon pan of the scientific literature. While other, lesser, theories become dirty and unusable when pelted with disconfirming data, these simply slide off the Geschwind theory, which continues to survive and inspire despite important critiques all pointing to the conclusion that the current status of the theory should be an-ambitious-idea-that-didn't-work-out.[29] For example, as the neurophysiologist Ruth Bleier pointed out more than two decades ago, the very starting point of the theory – the idea that the foetal male's higher level of testosterone brings about a more cramped left hemisphere – was inconsistent with a large postmortem study of foetal brains.[30] More recently, a neuroimaging study of seventy-four newborns also found no evidence of a relatively smaller left hemisphere in males.[31]

But still, the idea that higher foetal testosterone somehow creates a 'male' brain that is superior in masculine things like science and maths, while lower foetal testosterone leads to a touchy-feely, 'female' brain, has tremendous appeal. Baron-Cohen's hypothesis is an elaboration of the Geschwind theory. His idea is that low levels of foetal testosterone result in a female, E-type brain; medium levels yield a balanced brain; and high levels of foetal testosterone make for a male, S-type brain. (And really high levels of foetal testosterone create an 'extreme male brain' that is good at systemising, really bad at empathising, and is also known as autistic.)[32] Since there is overlap between the sexes in foetal-testosterone levels in the second trimester – some girls have higher levels than some boys – this would explain why some females are systemisers and some males are empathisers. But because, on average, males have higher testosterone levels, they will be more likely to have S-type brains. That's the idea: how do we test it? It's not that easy. Higher levels of foetal testosterone are strongly correlated with having a penis. That means that a correlation between foetal-testosterone levels and later sex-typed behaviour, or differences between boys and girls, could have nothing to do with foetal testosterone and everything to do with the different socialisation of boys and girls. But as we'll see in the next two chapters, there are several ways around this problem.

What will they tell us about the biological basis of gender inequality?

> Without testosterone interfering, your daughter developed not only female genitalia but a decidedly female brain . . . it is your daughter's girl brain that will direct her female approach to the world.
>
> —the Gurian Institute, *It's a Baby Girl!* (2009)[1]

At this point in the book, you may have begun to be a bit suspicious of phrases like 'female approach to the world.' As we discovered earlier, a person's approach to the world can depend on what kind of social identity is in place or the social expectations that are salient. The girl brain directs not so much a female approach to the world as a flexible, context-sensitive one. But that's not to say that foetal testosterone isn't doing *something* in the brain. And perhaps the most obvious strategy for working out what that might be is to compare the empathising and systemising skills of children and adults who were exposed to different levels of foetal testosterone. If girls with higher foetal testosterone are more masculine than girls with lower levels (and ditto for boys), then this could mean that children with higher foetal testosterone have brains that have been more 'masculinised' in utero. (Then again, it might not.)[2]

One technical difficulty with this approach, however, is that only extremely rarely is blood sampled from an unborn baby. This means that researchers can't directly measure the amount of testosterone circulating in the baby's blood. So what do they do instead? Some researchers measure maternal testosterone, the testosterone level in the blood of the pregnant mother. Other researchers measure the amniotic testosterone in the amniotic fluid (which is taken from the sac surrounding the foetus for the purposes of prenatal testing). Yet other researchers study adults and use digit ratio as a proxy for the foetal testosterone levels. The 2D:4D digit ratio is the ratio of the length of the second (index) finger and the fourth (ring) finger. This ratio is, on average, different in men and women. (Men tend to have longer ring fingers relative to their index fingers, while women's index fingers are about the same length as, or slightly longer than, their ring fingers.) The idea is that prenatal testosterone levels influence digit ratio. These very different approaches all have something very important in common: researchers don't actually know for sure whether what they are measuring correlates well, *or even at all*, with the level of testosterone acting on the foetal brain.[3] We won't let this hold us back. (After all, we're only trying to find the biological roots to gender inequality, so why be fussy, right?) But it's worth bearing in mind.

With all the nitpicking done, we're ready to look at the evidence that the 'female approach to the world' begins in the womb.[4] In a series of articles, Simon Baron-Cohen and his colleagues have described a large group of children whose mothers had amniocentesis in the second trimester of pregnancy. According to his hypothesis, higher amniotic testosterone should bring about worse empathising skills. So, does amniotic testosterone negatively correlate, in boys and girls separately,[5] with frequency of eye contact at twelve months old with a parent during play, quality of social relationships at four years old (as assessed by the mother), propensity to use mental-state terms, scores on the child version of the Empathy Quotient (EQ; as assessed by the mother), and performance on a child's version of the Reading the Mind in the Eyes test? The answers are, respectively: no;[6] not really;[7] not really;[8] no;[9] and yes.[10] And before you get too excited about this last yes for the Reading the Mind in the Eyes test, even though performance correlated with amniotic testosterone, girls scored no better than boys.[11] Expanding the scope of the search to include digit-ratio studies also yields little in the way of support.[12]

What about prenatal testosterone and systemising? Systemising, you will recall, is 'the drive to analyze or construct systems', and '[a] system is defined as something that takes inputs, which can then be operated on in variable ways, to deliver different outputs in a rule-governed way.'[13] As the observant reader might have noticed, we have yet to encounter an actual test of systemising ability. Nor can we even assume that a strongly systemising brain is the best kind to have to become a top-notch scientist. Philosopher Neil Levy has suggested that '[i]ntelligence, even in the hard sciences, and even in innovation, is as much an "empathizing" power as it is systemizing.' Albert Einstein, for example, described his breakthroughs as being the result of 'intuition, supported by being sympathetically in touch with experience' rather than the end point of a 'logical path'.[14] Nobel Prize winners agree. An analysis of the transcripts of interviews with these illustrious men and women of science found that the majority accept that there is such a thing as scientific intuition that is distinct from conscious, logical reasoning and that can take place in the absence of all the information necessary for logical reasoning. In fact, their descriptions of scientific intuition bear a striking resemblance to Baron-Cohen's characterisation of empathising as 'an imaginative leap in the dark in the absence of complete data'.[15] As one Nobel Prize winner in chemistry put it, 'Intuition, I always feel, is when we don't have enough components and yet we have to construct a picture.' And while of course logical reasoning is vital, this intuitive scientific process that many laureates described as helpful to them can be undermined if this is the only approach taken, as a laureate of medicine describes:

> This apparatus . . . which intuits has to have an enormous basis
> of known facts at its disposal with which to play. And it plays
> in a very mysterious manner, because . . . it sort of keeps all

known facts afloat, waiting for them to fall in place, like a jig-saw puzzle. And if you press . . . if you try to permutate your knowledge, nothing comes out of it. You must give a sort of mysterious pressure, and then rest, and suddenly BING . . . the solution comes.[16]

This is another point to bear in mind when we consider the strength of the evidence for prenatal origins to gender inequality in science. In truth, '[n]o perfect set of cognitive abilities that makes one a successful scientist has been identified'.[17] (Needless to say, this makes the task of finding the prenatal origins of such success that much harder.)

But let's just accept the assumption that systemising is an important key to success in science, and return to the data. A study from Simon Baron-Cohen's lab looked for, and found, correlations between amniotic testosterone and something promisingly named the Systemizing Quotient (SQ) for children (filled in by the mother).[18] Yet while some of the items on this questionnaire have a systemising-y feel to them (asking, for example, whether the child can 'easily figure out the controls of the video or DVD player' or 'knows how to mix paints to produce different colours'), for many other questions one struggles to understand how they tap into a desire to understand input-operation outputs. In what way does minding 'if things in the house are not in their proper place', becoming 'annoyed when things aren't done on time', or noticing 'if something in the house had been moved or changed' reflect a mind driven to understand the rules of the law-bound universe?[19] I'm not the expert here, but I can't help wondering if some of the items from the Fusspot Quotient accidentally found their way into the SQ.

Slightly more on target is a study of the toy choices of thirteen-month-old children. Boys spent more time than did girls playing with the boyish toys, which were a trailer with four cars, a garbage truck, and what was somewhat unhelpfully described as 'a set of three plastic pieces of equipment'. Are these systemising toys? I suppose you could make a case for it. You push a car or a trailer, it moves. And we'll give the 'plastic pieces of equipment' the benefit of the doubt. Certainly, these toys are probably better candidates than the tea set, dolls, baby bottle and cradle with which girls spent more time than boys. But then again, the three gender-neutral toys (a plastic friction dog, a wooden puzzle and a stacking pole with rings), with which boys and girls spent equal time, seem at least as systemising as the boyish toys, if not more so. Not that it matters, since neither amniotic testosterone nor maternal testosterone turned out to be related to play behaviour anyway.[20] (Disclaimer: When I say 'boyish' toys, I am referring to toys traditionally marketed to boys; likewise for 'girlish' toys.)

Nor do studies of correlations between amniotic testosterone and cognitive performance lend much support to the idea that higher prenatal testosterone is associated with greater skill on visuospatial tasks, mathematics, or other vaguely scientific-like skills. Does accuracy on a mental rotation test at age seven correlate with amniotic testosterone? No.[21] Does a four-year-old's skill at copying a block structure, understanding number facts and concepts, and counting and sorting increase with higher levels of amniotic testosterone? No, it *decreases* in girls, and has no relationship in boys. Puzzle solving? No. Classification skills (for example, 'find all the small ones'?) No.[22] A test of spatial ability? No.[23] And again, while some digit-ratio studies do provide a spattering of support, others have failed to find correlations between digit ratio and SQ score, and mental rotation ability. One study even found that physical scientists have more-feminine digit ratios than do social scientists.[24] There are a few more prenatal testosterone studies, which we'll come to in a later chapter. But there is, I think, something a little underwhelming about the evidence so far.

The prenatal-testosterone studies are, however, just one source of evidence for the fetal fork hypothesis. The period shortly after the baby is born supposedly provides another:

One of the first things your daughter's female brain will compel her to do is study faces. Whereas child developmental specialists originally thought all infants came wired for mutual gazing, your daughter may be more interested in staring at a human face than the newborn male.[25]

This quote from the Gurian Institute's book *It's a Baby Girl!* is a typical popular take on a study conducted several years ago by Simon Baron-Cohen, together with graduate student Jennifer Connellan and other colleagues. They looked for gender differences in newborns who were on average just a day-and-a-half old. The logic was simple: any differences between the sexes seen at this tender age can't be chalked up to socialisation. One hundred and two babies were offered, one at a time, Connellan's own face and a mobile to look at. The idea was to measure the babies' interest in the face versus interest in the mobile: empathising versus systemising. Each baby's eye gaze was filmed, and this recording was later used to time how long each baby spent looking at the face and the mobile. Male and female babies spent equal amounts of time looking at the face: both sexes, on average, spent just under half the total looking time (which was about a minute) looking at Connellan's face. However, males looked longer at the mobile than did females (51 percent of looking time versus 41 percent for females) and females, as a group, looked longer at the face than the mobile (49 percent versus 41 percent of looking time).[26]

Much has been made of the significance of this study. 'The results of this experiment suggest that girls are born prewired to be interested in faces while boys are prewired to be more interested in moving objects', writes Leonard Sax in his book *Why Gender Matters*,[27] a conclusion echoed in the popular media around the world. The implications for career choices are clear. Cambridge academic Peter Lawrence, citing the newborn study, argues that men and women are 'constitutionally different' and thus unlikely to ever become professors of physics and literature in equal numbers.[28] And in his contribution to the book *Why Aren't More Women in Science?* Baron-Cohen suggests from the newborn study that the '"bias" in attention to things rather than emotions (in boys) and vice versa (in girls)' reflects 'partly innate differences' that culture then amplifies. Sex differences in the empathising versus systemising bias, Baron-Cohen argues, 'suggests that we should not expect the sex ratio in occupations such as math or physics to ever be 50-50 if we leave the workplace to simply reflect the numbers of applicants of each sex who are drawn to such fields.'[29] In other words, short of some very heavy-handed social engineering, gender equality in the workplace is an impossible ideal.

But unfortunately, as some researchers have pointed out, the study was simply not done well.[30] When you are claiming nothing less than evidence of the biological origins of a gender-stratified society, it helps to have a methodology that stands up to scrutiny. No study is perfect, of course, but this one was flawed in ways it simply need not have been, as psychologists Alison Nash and Giordana Grossi have pointed out. Some of these problems concern the sort of detail that may provoke a small yawn in the nonspecialist, but a severe case of eyebrow-in-the-hairline for experts. First of all, there are standard procedures when it comes to testing newborns for their visual preferences. A baby's attention span is not at its peak in the first few days of life, waxing and waning over short periods of time. For this reason, when infant researchers want to find out which of two stimuli a newborn finds most interesting, they usually present them simultaneously. If you don't, and instead present them one after another, then you don't really know whether the baby looked at stimulus A more because she genuinely found it more interesting, or whether she was irritated by some internal rumblings, about to fall asleep, or simply a little tired of life when stimulus B was on show.

In Connellan's study, the face and the mobile were presented separately.

Another important thing to know about very little babies is that they can't see very well. They actually aren't even drawn to faces per se but to visual stimuli that, like the face, have a top-heavy pattern. In fact, before the age of three months, babies actually prefer top-heavy, facelike patterns over real faces. It's important, therefore, to ensure that babies all view the stimuli from the same angle, otherwise the same stimulus can appear to be different, including its degree of top-heaviness.

In Connellan's study, some babies were tested on their backs in a cot, and other babies were tested in a parent's lap. (If more girls than boys, say, were on their backs, then on average your groups of boys and girls are not seeing the same stimuli.)

But the most major problem with the study, described by Nash and Grossi as a 'striking design flaw', was its potential for experimenter expectancy effects.[31] If you have ever visited a new mother in a maternity ward, there is a good chance that you will have seen one or more of the following items: a baby wearing a pink or blue (or otherwise gendered) outfit; a pink or blue balloon; a pink or blue blanket; an arrangement of predominantly pink or blue flowers; pink or blue congratulation cards; or even (as was the case in the hospital in which I gave birth) a pink or blue name card on the baby's bassinet. Clues, in short, as to the baby's sex. Now if you are an experimenter and stimulus rolled into one neat package with a particular hypothesis in mind (not to mention a head full of cultural assumptions), you have to make sure that this information doesn't unconsciously affect your behaviour towards the baby. This, of course, is impossible. As we saw in the first part of this book, even information that doesn't register with consciousness can subtly change behaviour. Researchers therefore usually take this problem very seriously and go to some effort to eliminate experimenter expectancy effects. Here, for example, are the precautions taken by another recent study that also looked for gender differences in newborn eye gaze:

> We instructed all participants that the infant must be dressed in a gender-neutral outfit and that the interacters in the study room must remain unaware of the baby's sex throughout the interaction, as well as after the interaction was complete. Because parents often had either pink or blue outfits for their

newborn, many opted to dress their baby in the white outfits provided by the hospital . . .

We decided that the study should take place in a room other than the mother's room in order to decrease the likelihood that something in the room would provide clues to the interacters as to the sex of the infant. . . .

To keep the interacters blind to the sex of the infant all identifying information on the infant's bassinet was covered or removed upon arrival to the study room.

Researchers do not go to such lengths merely to make life awkward for themselves and the parents of newly born babies. (In this carefully designed study, no gender differences in eye gaze were found in newborns although, interestingly, they did find gender differences in eye gaze in a follow-up three to four months later. This, they point out, suggests the possibility 'that the gender-typed behaviour pattern is not innate but, instead, learned in early infancy.')[32]

No such precautions were taken in Connellan's study.

She knew the sex of at least some of the newborns she tested, and it's not beyond the realm of possibility that, on other occasions, clues as to the baby's sex unconsciously undetected could have swayed her behaviour in a direction consistent with gender stereotypes.[33] Unfortunately, this was a study in which even slight differences in the experimenter's behaviour could well create experimenter expectancy effects. Motion, open eyes and mutual eye gaze are all visual stimuli that newborns especially like and are sensitive to.[34] It is, I imagine, rather hard to hold up a mobile, and look at a newborn, in exactly the same way 102 times. What if Connellan inadvertently moved the mobile more when she held it up for boys, or looked more directly, or with wider eyes, for the girls?

But even if a redoing of the study, performed with a less cavalier approach to normal policy and procedure in infant testing, got the same result, what would it actually signify? Nash and

Grossi have argued that if the sex differences in the newborn study reflect differences in brain organisation then we should see increasing divergence between girls and boys as these skills develop. Yet boys' greater interest in the mobile doesn't seem to serve them much advantage. As Nash and Grossi have pointed out, as has Harvard University developmental psychologist Elizabeth Spelke, there is little evidence for a systemising advantage in young boys: a large body of research exploring infants' understanding of objects and mechanical motion finds no advantage for males.[35] As for the development of empathy, evidence of divergence is modest. Boys and girls develop an understanding of the mental states of others at a similar rate. But girls do have a small advantage, on average, in facial expression processing and, overall, studies find signs of greater affective empathy in girls. However, as is the case in adults, this difference is much smaller when based on observations rather than self-report or report by another (such as a parent).[36] But also, these psychologists have pointed out, why think that what a newborn prefers to look at provides any kind of window, however grimy, into their future abilities and interests? It might come down to something as boring as girls responding more or less to some other difference between the two kinds of stimuli – visual, auditory or olfactory – that has nothing to do with faces versus objects per se. We have no idea whether newborn preferences reflect what their later abilities will be – such an assumption is, as Neil Levy puts it, 'essentially unargued for' and 'questionable at best'.[37]

Many studies have methodological flaws. Many studies are overinterpreted. But not many studies inspire in their authors and others the conclusion that innate differences in part lie behind our gender-stratified society.[38] This is a study that really needs to be repeated before it is taken too seriously, and with closer attention to what the results might actually mean, as well as those little details that make all the difference between the study the expert feels she can trust and the study that leaves her eyebrow muscles aching and exhausted.

So what *does* go on in the darkness of the womb? Consider the boldness of the statements made in the popular media about the effect of foetal testosterone on the brain. Now consider the inadequacy of the data showing links between exposure of the foetal brain to testosterone (which, you will recall, these studies might not even be tapping) and brain 'type'. Contrast, for a moment, the confidence of claims that boys and girls arrive with differently prewired interests, against the flimsiness of the evidence. There's something a little shocking about the discrepancy between the weakness of the scientific data on the one hand and the strength of the popular claims on the other. As Simon Baron-Cohen himself has written, 'the field of sex differences in mind needs to proceed in a fashion that is sensitive . . . by cautiously looking at the evidence and being careful not to overstate what can be concluded.'[39]

At last, something on which we can all agree.

It's a good life. If I die tomorrow, I'll die a happy woman, because
I'll feel like I've done a lot of good work.

—Kerin Fielding, orthopaedic surgeon[1]

Today, women are strongly represented in fields such as biology, psychology, medicine, and forensic and veterinary science. Some think this reflects 'the feminine propensity to protect and nurture – and the desire to work with living things', as Christina Hoff Sommers suggested by way of explaining the recent influx of women into the once male-dominated domain of veterinary medicine.[2]

Maybe. But there is something a little unsatisfying about this reframing of the life sciences as: *Now with added empathising for extra feminine appeal!* Is the supposed female drive to work with living things, or to engage with mental states, really likely to be satisfied by looking at cells under microscopes or de-sexing cats? Even academic psychology, most of which *is* at least about people, is devoted to the pursuit of understanding the laws and principles – one might even say *systems* – that underlie cognition and behaviour. Apart from the lab teamwork common to science in general, the core work of an academic psychologist – making sense of the literature, designing experiments and analysing and interpreting data – puts few demands on empathising abilities. And what about forensic science, which draws in more than three times as many women as men?[3] On the one hand, it does indeed sometimes have *people* as its subject of study. But, on the other hand, when it does, often they are dead.

As journalist Amanda Schaffer has pointed out:

[I]f history is any guide, today's gender breakdowns are likely to keep changing. What's so magical, after all, about the current numbers? A few decades ago, most biology and math majors were men. So were most doctors. Now maths undergraduate majors split close to 50/50. In 1976, only 8 percent of Ph.D.s in biology went to women; by 2004, 44 percent did. Today, half of M.D.s go to women. Even in engineering, physics, chemistry, and math, the number of women receiving doctorates tripled or quadrupled between 1976 and 2001. Why assume that we have just now reached some natural limit?[4]

It's a good point. Perhaps in a few decades we will be redefining women's new levels of participation in the physical sciences, politics and business as reflecting their innate drive to nurture. After all, is there any more powerful way to help others than to develop sustainable technologies, set tough emissions targets or, like Bill Gates, write big fat cheques to charitable causes?

As some psychologists have pointed out, such historical shifts – including the movement away from male dominance in teaching and secretarial work – don't lend themselves especially well to explanations in terms of hormones and genes.[5] So with this malleability of sex segregation in mind, let's turn to the next two ways of investigating the link between foetal testosterone and later sex-typed behaviour: females whose in utero living conditions were, hormonally speaking, wrong for their chromosomal sex; and monkeys.

In a condition called congenital adrenal hyperplasia (CAH), the child's genetic state results in the foetus's being exposed to unusually high levels of testosterone. In girls with CAH, this triggers development of male external genitalia. (The female internal reproductive organs, however, develop normally.) Girls with CAH are born with genital virilisation – that is, they look more-or-less like a boy at birth, depending on the severity of the condition. Usually the condition is detected at birth. The child is then given ongoing hormonal treatment, some time later undergoes surgery to feminise her genitalia, and is raised as a girl. This offers an opportunity for researchers to explore the effects of high foetal testosterone, disentangled from what normally comes with that experience, namely, also being reared as a boy. However, it's important to point out that girls with CAH are not simply girls plus extra foetal testosterone. Not only are other hormone levels also awry (and are therefore potential candidates for being behind any differences in behaviour), but also these girls are born with ambiguous genitalia, and receive continuous hormonal treatment as well as, most likely, extensive surgery on the genitalia. (When this happens seems to be quite variable.) It's not impossible to imagine that this could create a certain ambivalence around the child's gender in the mind of a parent, and perhaps in the child herself, for which there is a little evidence.[6]

But, nonetheless, are girls with CAH more likely to be systemisers than empathisers? So far, we can't say. Older girls and adults with CAH do report less tender-mindedness, interest in infants, and social skills than their non-CAH relatives. But on the other hand, they report equal communication ability (assessed with questions like *I am good at social chit-chat*, and *I find it easy to 'read between the lines' when someone is talking to me*) and no greater dominance (which includes masculine qualities like being aggressive, authoritative and competitive).[7] So the evidence is a little mixed and, as we learned in Chapter 2, self-report scales may tell us little about people's actual empathic tendencies and skills. As for systemising, in the absence of an actual test of this ability it's impossible to know. One study found that girls with CAH report *less* attention to detail than control girls (a skill that Baron-Cohen considers especially important for systemising).[8] And there's no evidence that the high prenatal-testosterone levels of CAH serve to improve mathematical performance – it's even been suggested that it *impairs* it.[9] Researchers have also tested girls with CAH on the ubiquitous mental rotation tasks, and the evidence currently points towards an advantage for them over unaffected girls.[10] But, as has been pointed out, this could be the result of their more boyish play experiences, rather than prenatal testosterone per se.

And girls with CAH definitely do differ from their non-CAH sisters and relatives in their play. In as much as we can take at face value their caregivers' reports and behaviour when under observation in the lab, this seems to be despite the best efforts of their parents.[11] Girls with CAH play much more at boyish activities and toys than do control girls (although not quite as much as boys do), and they are also less interested in girlish toys and pastimes.[12] This boyishness seems to continue into adolescence. For example, adolescent girls with CAH are intermediate between boys and girls in their interest in sex-typical activities (football versus needlepoint, embroidery or macramé) and future occupations (like engineer versus professional ice skater).[13]

These tomboyish interests seem to provide a compelling case for the idea that foetal testosterone organises the brain to be drawn to certain kinds of stimuli that lie behind sex differences in play behaviour and, by implication, occupational segregation.[14] But what is a little odd is that no attempt seems to have been made to work out whether girls with CAH are drawn to some particular *quality* in boyish toys and activities or whether they are drawn to them simply by virtue of the fact that they are associated with males.[15] Take, for instance, the Pre-School Activities Inventory, on which girls with CAH score more like males than unaffected girls. The inventory includes questions about playing with cars and dolls, and so on.[16] But girls with CAH can also get a higher score than unaffected girls by, for example, showing little interest in jewellery, pretty things, dressing up in girlish clothes and pretending to be a female character.[17] Another study (drawing on a different clinical group) found that greater prenatal androgen exposure led to less interest in activities like ballet, dressing up as a fairy, dressing up as a witch, dressing up as a woman, gymnastics, playing hairdresser and working with clay, but more interest in basketball, dressing up as an alien, dressing up as a cowboy, dressing up as a man, dressing up as a pirate and playing spaceman.[18] Likewise, women with CAH asked to recall their childhood activities score significantly differently from controls on a questionnaire that, among other questions, asks about use of cosmetics and jewellery, hating feminine clothes, the gender of admired or imitated characters on TV or in movies and whether they dressed up more as male or female characters.[19]

In most lab-based toy studies, too, there is a question mark over what the researchers are really measuring. The boyish toys on offer always include vehicles and construction toys, while the girlish toys always include dolls with accessories and tea sets. (Interestingly, one of the staples of the boyish toys, the Lincoln Logs construction set, recently had to be replaced because girls liked it so much!)[20] But if it's stimulation of their visuospatial skills that girls with CAH are drawn to, why don't they (and boys, for that matter) spend longer than girls on the neutral toys, which often include a puzzle and a sketchpad? What form of brain masculinisation could lead to a preference for dressing up as an alien rather than a witch, an interest in fishing over needlepoint, a desire to wash and wax the car rather than try out for cheerleading, or masculine costumes over feminine ones?[21] Is it possible that what researchers are seeing in girls with CAH is greater identification with male activities, whatever they might be?

Interestingly, studies that have looked at the correlations between early testosterone and later gendered-play behaviour in nonclinical children – which so far have shown the most convincing relationships (although they are still not very impressive) – encounter this very same problem. For example, one study found correlations between amniotic testosterone and male-typical play within both boys and girls, while an earlier study found a correlation between maternal testosterone and play behaviour, although only in girls. But in both studies the behavioural measure used was the Pre-School Activities Inventory, which, as mentioned earlier, includes items that may have more to do with cultural gender rules than more fundamental psychological predispositions. (A third study, using a different measure of gendered play, found no relationship at all between amniotic testosterone and play preferences.)[22]

In short, we just don't know what's going on. One researcher has suggested that 'androgen may affect the reward value of moving stimuli, so that objects that move and have moving parts may be more rewarding to girls with CAH and to boys than to typical girls.'[23] But we just don't know until this idea is tested. If in these toy preference studies Barbie came with a pink car instead of clothes and hair accessories, would girls with CAH play with her more than control girls? That's what the brain organisation hypothesis would predict. Would a girl with CAH rather play with a toy stroller that can be wheeled around, over a firetruck that cannot? Would the changing proportion of men in an occupation, like veterinary medicine, have no effect on its appeal to girls with CAH?

Perhaps. But another possibility is that girls with CAH are drawn to what is culturally ascribed to males. Thirty years ago, primatologist Frances Burton put forward an intriguing suggestion that casts the data from females with CAH in an entirely new light. She proposed that the effect of foetal hormones in primates is to predispose them to be receptive to whatever behaviours happen to go with their own sex in the particular society into which they are born.[24] (We'll shortly see what led her to this hypothesis.) As Melissa Hines points out, this would provide a very 'flexible design', enabling 'new members of the species to develop sex-appropriate behaviors despite changes in what those behaviors might be. This hormonal mechanism would liberate the species from a "hard-wired" masculinity or femininity that would be unable to adapt to changes in the environment that make it advantageous for males and females to modify their niche in society.'[25]

However, Hines has argued that this can't be the whole answer to gender differences in toy preferences. This is because, remarkably, similar sex differences in toy preference are also seen in monkeys. In a study with Gerianne Alexander, Hines put six toys, one at a time, into a large enclosure of vervet monkeys. There were two boyish toys (a police car and a ball), two girlish toys (a doll and a pan) and two neutral toys (a picture book and a stuffed dog). They measured how long each monkey spent with each toy, as a percentage of total toy-contact time. Both male and female vervets spent about a third of the total time with the neutral toys. Male vervets spent about another third each of their total playing time with the other toys. By contrast, females spent more time with the girlish toys than with the boyish toys.[26] If, by the way, you are curious about the choice of a pan as a girlish toy, you are not alone. Although it is true that primatologists regularly uncover hitherto unknown skills in our nonhuman cousins, the art of heated cuisine is not yet one of them. Frances Burton has informed me that, in her long career of observing monkeys, she has never met one that could cook.[27] (This raises the more general point, spontaneously made by more than one of the academics who read this chapter, that it is not at all clear that a toy taken from human culture has the same meaning to a monkey, to which it is unfamiliar, that it does to a child.)[28] It's worth noting, then, that when the researchers divided up their stimuli in a different way – comparing amount of play with animate toys (the dog and the doll) with object toys (the pan, ball, car, and book) – they found no differences between the sexes.

After an interval of about six years, a second group of researchers ran another toy-preference study with rhesus monkeys. This study was different in two important ways. First of all, trying to get to the bottom of *why* there are gender differences in toy preference, they compared wheeled toys that invite movement with stuffed-animal toys that supposedly invite nurturing. (Whether or not the stuffed animals *were* actually nurtured is unclear, especially as one trial had to be terminated early when 'a plush toy was torn into multiple pieces'.) Second, the researchers gave monkeys an outright choice between the two types of toy – one of each was put into the enclosure at the same time, which is a better test of preference. They found that females were as interested in wheeled toys as they were in plush ones, and played no less with wheeled toys than did male monkeys. However, unlike females, male monkeys had a preference for wheeled toys over plush ones.[29]

What are we to make of the subtle sex differences seen in these two slightly contradictory studies? (Which doesn't seem like quite large enough a number on which to base any terribly firm conclusions about human nature.) One reasonable summary might be that male and female monkeys alike enjoy playing with both stuffed toys and mobile objects, but that in males the cuddly dolls have less of a shine than the mobile toys. (Just to confuse matters, stuffed toys don't seem to be disfavoured by either vervet males or boys.)[30] What does this mean for humans, and the toys played with by little boys and girls?

These two studies have been taken as strengthening the evidence of 'inborn influences on sex-typed toy preferences',[31] support for the idea that 'biologically based sex differences in activity preferences significantly influence sex differences in childhood object choice',[32] and 'another nail in the coffin for the idea that similar preferences in human children are entirely due to culture'.[33] Yet can we safely move to the conclusion that the higher levels of prenatal testosterone normally seen only in males increases interest in boyish toys that move or stimulate visuospatial skills, and reduces interest in toys related to babies and nurturing? These are two separate effects that are hard to disentangle when you compare interest in a moveable boyish toy *relative* to interest in a nurture-able girlish toy. Although male rhesus monkeys preferred the wheeled toys over the plush ones, because there was no gender-neutral toy condition we don't really know whether rhesus males were especially drawn to the wheeled toys or simply *less* interested in the plush animals. After all, in the first monkey study male vervets spent no longer with the moveable ball and car than with the neutral toys or the girlish toys. So neither monkey study does a convincing job of showing that male monkeys are born with a built-in interest in objects that move. Researchers need to get more specific about what particular feature of boyish toys supposedly appeals to the male brain, and then see whether male monkeys more than females prefer novel toys that do have this feature over other equally novel toys that don't.

But what about the idea that females, thanks to their lower foetal-testosterone levels, are born with a greater built-in interest in toys that lend themselves to nurturing play? It's a compelling interpretation, especially given the lack of interest in babies and dolls shown by girls with CAH. (Interestingly, they are no less interested in pets.)[34] The only problem is, prenatal-testosterone levels have been found to have *no effect* on male or female rhesus monkeys' interest in infants. Male youngsters whose mothers had been experimentally treated prenatally with an androgen-receptor blocker were no more interested in infants than control males, despite their more-feminised hormonal environment. And crucially, female youngsters whose mothers had been given testosterone injections during pregnancy were no *less* interested in infants than control females. It should be said that the researchers who reported these surprising results, seeing no evidence that mothers differentially socialised male and female infants, declared themselves 'reluctant . . . to dismiss prenatal hormonal influences altogether' in explaining sex differences in interest in infants among rhesus monkeys.[35] Yet there is good reason to think that this reluctance may be misplaced.

Frances Burton has pointed out that, just like us, primate societies have norms regarding which sex does what: who gets food, rears the young, moves the troop, protects the troop and maintains group cohesion.[36] But, these norms are different across, or even within, primate species. Male involvement in infant rearing, for instance, ranges from the hands-off to the intimate. For example, 'a specially intimate relation between adult males and infants' has been seen in some troops of wild Japanese macaque monkeys (the species *Macaca fuscata fuscata*) during delivery season: males protect, carry and groom one- and two-year-old infants. Yet different troops of the same species, in different parts of the country, show less of this paternal care, or even none at all.[37] Similarly, in another species of macaque (*Macaca sylvanus*) Burton has seen extensive and lengthy male care of young in a Gibraltar troop. Indeed, so important is male baby-sitting in this troop that 'young females are kept away from infants so that young males may learn their role.'[38] Yet among the very same species in Morocco, male care is much less significant.

As Burton argued, 'while hormones are the same' throughout these different species, there is 'no universal pattern' to how the different tasks of the society, including infant care, are divided. Sometimes both sexes perform the role, sometimes only one or the other sex does. 'If the hormones determine the roles, one would expect to find the same sex occupying the same roles in all societies. This is patently not the case'.[39] In line with this flexibility, it seems that the potential for primate male care-giving is by no means destroyed or even diminished by foetal testosterone. Another primatologist, William Mason, points out that 'schemas for parental behaviour are present in infancy, they appear in the same form in both sexes, and they continue to be accessible throughout life.'[40] However, interest towards infants soon begins to diverge in the sexes. At one year of age, male and female rhesus monkeys exhibit few differences in behaviour towards infants. Yet at two and three years of age, females contact, embrace, groom, touch and initiate closeness with infants more often than do males – and the females who show this greater interest in infants include females treated with prenatal androgens.[41] We may need to look elsewhere to find a reason for the lack of interest in infants and dolls in girls with CAH.

So how does a male macaque monkey in Takasakiyama, Japan, become an involved carer while his counterpart in Katuyama perfects paternal indifference?[42] Perhaps the action of prenatal testosterone on the genitalia plays an important part in explaining how primate infants come to learn the idiosyncratic traditions of their group. Monkeys take great interest in the genitalia of new-borns. Unable to avail themselves of the convenience of observing whether it is a pink or blue balloon tied to the entrance of the nest, monkeys take a more direct approach to satisfying themselves as to the answer to the question that appears to be as important to them as it is to us:

> In most monkey societies, the neonate is a strong attraction: all members of the troop rush over; attempt: to touch or hold it, sniff it, lick it, and otherwise exhibit interest in it. Through visual and olfactory stimuli, the sex of the individual is as much registered as its maternity.[43]

Is this interest in genitalia purely academic? To suggest that non-human primates have socially constructed gender roles seems more or less akin to pinning a notice to one's back that says, MOCK ME. But does the registration of sex – of others and perhaps of self – play an important role in maintaining traditional sex-division of labour in primate societies? When Burton studied troops of macaque monkeys in Gibraltar, she observed that the head male was intimately involved in neonate care: sniffing, licking, caress-ing, patting, holding and chattering to it, as well as encouraging it to walk. Interestingly, when the head male was in charge of the infant, he would be followed and imitated by subadults – but only males. The male subadults then themselves became involved in caring for the infant.[44] As we'll see in the third part of the book, human children have a powerful drive to self-socialise into gender roles. That is, even in the absence of any encouragement by par-ents, they are attracted to things and behaviours associated with their sex. Although children from the age of about two have the advantage of an explicit, reportable knowledge of their own sex, is it possible that some primitive sense of sex identity brings about self-socialisation in nonhuman primates? As Hines and Alexander recently asked, 'if some animals of one sex could be trained to use a particular object, would others of that sex model them?'[45]

If more researchers interested in human gender differences start to investigate questions like this, which acknowledge that nonhuman primates, like us, have social norms that need to be learned, perhaps the answers will surprise us.

For many years, attention was focused on adulthood sex differences in the levels of hormones like testosterone and oestrogen. Could these circulating sex hormones, via their effect on cognition, go some way towards explaining gender inequality? Many assumed too quickly that it did. Unfortunately, as Hines concludes from her review of this research, 'influences have been assumed to exist despite a lack of consistent supporting data.'[46] To offer just one comical example, various studies have found that higher testosterone levels are associated with better mental rotation performance, worse mental rotation performance or equal mental rotation performance.[47] Likewise, Steven Pinker describes this literature as 'messy' and 'contradictory' (although he nonetheless thinks that 'something will be salvaged' from it).[48]

And so it seems as though foetal testosterone has become the explanation of choice for gender inequality in science. In a 2005 conference on diversifying the science and engineering workforce, Lawrence Summers, then president of Harvard University, controversially suggested that women might be intrinsically less capable, on average, of high-level science. Foetal testosterone was rushed to the scene of the mishap. In the *New Republic*, Steven Pinker reminded an irrationally outraged public that variations in sex hormones, 'especially before birth, can exaggerate or minimize the typical male and female patterns in cognition and personality.'[49] In the *New York Times*, Simon Baron-Cohen set out a path that passes from foetal-testosterone levels, to different brains, to different cognitive talents. He also cited Connellan's newborn study, in which

boys looked longer at a mobile, as support for Summers's suggestion that sex differences in science-related skills are innate.[50] And
Canadian researcher Doreen Kimura wrote in the *Vancouver Sun*
that Larry Summers was not mistaken in his suggestion that men
and women differ in their innate talents, because sex differences
'in levels of sex hormones early in prenatal life . . . strongly influence many behaviours into adulthood. Those behaviours include
the intellectual or cognitive pattern, hormonal influences being
especially well-documented for certain kinds of spatial ability, like
being able to mentally rotate or manipulate visual objects.'[51]

And yet as we've seen, higher foetal testosterone in nonclinical
populations has not been convincingly linked with better mental
rotation ability, systemising ability, mathematical ability, scientific
ability or worse mind reading. Connellan's newborn study was
gravely flawed. And the research with girls with CAH and non-
human primates – which at first glance seems to show that there
are built-in sex differences in toy preferences – turns out to jumble
up vague, untested ideas about what the male and female brain
might be interested in with what is socially ascribed to the two
sexes. One can't help but feel a weary sense of irony in response to
Pinker's complaint that the 'taboo' of innate sex differences 'need-
lessly puts a laudable cause [the modern women's movement] on a
collision course with the findings of science'.[52] So far as I can tell,
that collision has yet to occur.

And there's still so much inequality to be explained! We need
to press on, into the brain itself.

In 1915, the illustrious neurologist Dr. Charles L. Dana set out in the *New York Times* his professional opinion vis-à-vis the wisdom of women's suffrage:

> There are some fundamental differences between the bony and the nervous structures of women and men. The brain stem of woman is relatively larger; the brain mantle and basal ganglia are smaller; the upper half of the spinal cord is smaller, the lower half, which controls the pelvis and limbs, is much larger. These are structural differences which underlie definite differences in the two sexes. I do not say that they will prevent a woman from voting, but they will prevent her from ever becoming a man, and they point the way to the fact that woman's efficiency lies in a special field and not that of political initiative or of judicial authority in a community's organisation. There may be an answer to this assertion, but no one can deny that the mean weight of the O.T. and C.S. in a man is 42 and in a woman 38, or that there is a significant difference in the pelvic girdle.[1]

The passage of time has not borne out Dr. Dana's promising idea that the neural circuitry involved in political initiative is located in the upper half of the spinal cord. Without even knowing where in the nervous system the 'O.T.' and the 'C.S.' are located, I am fairly confident that judicial savvy does not lie in the extra four units of them bequeathed to men. But, at the time, this argument seemed plausible enough to be published in the *New York Times*. And who knows, perhaps it served to sway, or at least reinforce, opinion on the controversial subject of votes for women.

Today, we can easily recognise the prejudice behind the implications Dana drew from his neurological observations. But even as one hypothesis falls (*'The connection between the spinal cord and the pelvis? You really think it involved in some important way?'*), another is there to take its place.

As an empirical endeavour, the neuroscience of sex differences began in earnest in the mid-nineteenth century. The findings of Victorian scientists and medical men of the day were 'a key source of . . . opposition' to women's suffrage and equal access to higher education, notes Yale University historian of science Cynthia Russett.[2] Certainly, as she documents, they improved on the ideas of their predecessors who presented evidence to argue, for example, that women's intellectual inferiority compared with white men could be seen in the angle of their faces. As asserted by a late-eighteenth-century expert in the measurement of facial verticality, 'The idea of stupidity is associated, even by the vulgar, with the elongation of the snout, which necessarily lowers the facial line.'

Women did not fare well in such assessments, and were reported to share with the 'primitive' and 'savage' races an unfortunate lack of facial verticality. It was not long, though, before this crude measure was jettisoned in favour of the more sophisticated cephalic index, namely, the ratio of skull length to skull breadth. The cephalic index was, for a while, thought to be a promising indicator of mental capacity, but was reluctantly abandoned when it became clear that the head shapes of 'inferior' social groups, including women, did not segregate neatly from those of 'superior' groups. It was later believed, as noted earlier, that women's intellectual inferiority stemmed from their smaller and lighter brains. And when it became unavoidably evident that one could be slight of brain but substantial of intellect (and vice versa), the hypothesis was reluctantly abandoned, and the brain searched more intimately for the neural correlates of female inferiority.[3]

The tape measures and weighing scales of the Victorian brain scientists have been supplanted by powerful neuroimaging technologies, but there is still a lesson to be learned from historical examples such as these. State-of-the-art brain scanners offer us unprecedented information about the structure and working of the brain. But don't forget that, once, wrapping a tape measure around the head was considered modern and sophisticated, and it's important not to fall into the same old traps. As we'll see in later chapters, although certain popular commentators make it seem effortlessly easy, the sheer complexity of the brain makes interpreting and understanding the meaning of any sex differences we find in the brain a very difficult task. But the first, and perhaps surprising, issue in sex differences research is that of knowing which differences are real and which, like the initially promising cephalic index, are flukes or spurious.

In the statistical jargon used in psychology, p refers to the probability that the difference you see between two groups (of introverts and extroverts, say, or males and females) could have occurred by chance. As a general rule, psychologists report a difference between two groups as 'significant' if the probability that it could have occurred by chance is 1 in 20, or less. The possibility of getting significant results by chance is a problem in any area of research, but it's particularly acute for sex differences research. Suppose, for example, you're a neuroscientist interested in what parts of the brain are involved in mind reading. You get fifteen participants into a scanner and ask them to guess the emotion of people in photographs. Since you have both males and females in your group, you run a quick check to ensure that the two groups' brains respond in the same way. They do. What do you do next? Most likely, you publish your results without mentioning gender at all in your report (except to note the number of male and female participants). What you don't do is publish your findings with the title 'No Sex Differences in Neural Circuitry Involved in Understanding Others' Minds'. This is perfectly reasonable. After all, you weren't looking for gender difference and there were only small numbers of each sex in your study. But remember that even if males and females, overall, respond the same way on a task, five percent of studies investigating this question will throw up a 'significant' difference between the sexes by chance. As Hines has explained, sex is 'easily assessed, routinely evaluated, and not always reported. Because it is more interesting to find a difference than to find no difference, the 19 failures to observe a difference between men and women go unreported, whereas the 1 in 20 finding of a difference is likely to be published.'[4] This contributes to the so-called file-drawer phenomenon, whereby studies that *do* find sex differences get published, but those that don't languish unpublished and unseen in a researcher's file drawer.

Neuroimaging studies of sex differences are certainly not exempt from this problem. It's important to realise that the patches of colour you see on brain scans don't actually show brain activity. Although it may seem as though fMRI and PET enable you to see a snapshot of the brain at work (or, as popular writers Allan and Barbara Pease claim, 'to see your brain operating live on a television screen'),[5] this simply isn't the case. 'Unfortunately, these pretty pictures hide the sausage factory', as one neurologist put it.[6] fMRI doesn't measure neuronal activity directly. Instead, it uses a proxy: changes in blood oxygen levels. (PET uses a radioactive tracer isotope, which attaches itself to glucose or water molecules, to indirectly track blood flow.) Busier neurons need more oxygen and (after an initial dip) active brain regions have higher levels of oxygenated blood, because blood flow to that area increases. The oxygen is carried by the haemoglobin in red blood cells, and haemoglobin has slightly different magnetic qualities depending on how much oxygen it's carrying. This creates a signal in the scanner (which pulses a magnetic field on and off). Neuroscientists then compare the difference in blood flow in brain regions during the task they're interested in, with blood flow during a control task or rest state. (Ideally, the control task involves everything the experimental task entails – button pressing, word reading and so on – except for the psychological process you're particularly interested in.) Researchers test for significant differences in blood flow in various locations of the brain regions during the two tasks, and if tests indicate that it *is* significant, a blob of colour is placed at the appropriate location on the picture of the brain.[7]

In other words, those coloured spots on the brain represent statistical significance at the end of several stages of complicated analysis – which means there's plenty of scope for spurious findings of sex differences in neuroimaging research. Many studies use both male and female participants. The researchers may well check for gender differences but, if none are found, make no mention of it in the published report. What's more, because imaging is so expensive, a small number of participants is the rule rather than the exception, and small neuroimaging studies may be especially unreliable, because nuisance variables (like breathing rate and caffeine intake, or even menstrual cycle in women) can dramatically change the imaging signal without having any effect on behaviour.[8]

Neuroimaging also brings with it the teething problems of a technology that's still in its infancy. There are healthy controversies in the neuroscientific community regarding how statistical analysis should best be done. There's nothing wrong with this in itself, of course. But it is a little disconcerting that neuroimagers are now finding that reported sex differences in brain activation haven't been put to adequate statistical testing, or can come and go depending on how the analysis is done, or can fail to generalise to a distinct but similar task within a second group of men and women, or that the kind of analyses used to establish sex differences in brain activation can also 'discover' brain activation differences between randomly created groups (matched on sex, performance and obvious demographic characteristics).[9] For all these reasons, it's critical not to place too much faith in a single study that shows sex differences but instead to look for a consistent pattern.

The importance of this becomes very clear when we consider the influence of the nonstick theory of Norman Geschwind and his colleagues who, you'll recall, suggested that high levels of foetal testosterone in males result in a left hemisphere that is underdeveloped relative to the right. This led to the idea that male brains are more lateralised (or specialised) than female brains, on average. That is, males tend to stick to their shrivelled left hemisphere when grunting monosyllables and use the roomier right hemisphere when processing visuospatial stimuli. By contrast, women's brains are supposedly less lateralised: during both language and visuospatial tasks, women tend to use both sides of the brain.

Now this is not regarded as an unimportant 'I say *to-may-to*, you say *to-mah-to*' sort of difference within the scientific community. A specialised, keep-it-local structure is supposedly what underpins male superiority on certain visuospatial tasks. By contrast, the more collaborative 'Left? Right? Hey, we're all in this together' approach of the female brain supposedly explains their superior verbal skills, because they can more easily integrate information processed in different parts of the brain. The other side of the coin, however, is a more cramped design for spatial processing. Purportedly, this is because there is more competition between verbal and spatial circuits in the female, bilateral brain, which also, supposedly, has a relatively thicker and more bulbous corpus callosum, which is the bundle of neurons that connects the two hemispheres. This superior corpus callosum (especially a part of it called the splenium) supposedly enables faster and more efficient cross-talk between the hemispheres.[10]

There is something a little curious about the relationship between (some, at least, in) the scientific community and the idea of greater male lateralisation. It is a bit like that of the wife who determinedly overlooks the plentiful signs that her husband is shifty, unreliable and worthless, while inflating the significance of occasional dependable behaviour. Even in the 1980s, researchers were pointing out major flaws and yet, as Ruth Bleier noted in 1986, even 'devastating criticisms by two leaders in the field of cognitive sex differences and lateralization have done nothing to stem the flood of research'.[11]

Neuroimaging has provided a new way for researchers to show their loyalty to the hypothesis. Yet as neuroscientist Iris Sommer and her colleagues have shown, despite the new frisson of excitement wrought by the introduction of new technology, the data are as faithless as ever. Sommer and her colleagues reviewed (twice) all functional imaging studies of language lateralisation in a meta-analysis. (A meta-analysis is a statistical technique for putting together all studies that have investigated a particular question, taking into account the size of the study, to get a more accurate overall picture of the empirical situation.) The first meta-analysis (in 2004) put together data from more than 800 participants, and the second, in 2008, included more than 2,000 participants. In both meta-analyses they found 'no significant sex difference in functional language lateralization'.[12] Interestingly, they also found that studies that found sex differences tended to have smaller sample sizes than those that didn't. As Sommer and colleagues suggest, this may be a sign that the file-drawer phenomenon is at work, with biased reporting of chance findings from smaller studies.

Sommer also looked at older ways of looking for sex differences in language lateralisation. The left hemisphere processes auditory input from the right ear, and vice versa. If men, more than women, tend to use just the left hemisphere for language, then they should find it relatively easier to process words fed into the left hemisphere via the right ear (a phenomenon known as the right-ear-advantage). But Sommer and colleagues' meta-analysis of these data, from nearly 4,000 participants, found no sex difference in the right-ear-advantage.[13] (Nor does the whopping dose of foetal testosterone experienced by girls with CAH seem to bring about a larger right-ear advantage.)[14] Another approach is to see how stroke damage to the left or right hemisphere affects the language abilities of male and female patients. While early studies found that men were more likely to suffer language problems (aphasia) after left-hemisphere damage, later and larger studies have not found this, including the Copenhagen aphasia study of more than 1,000 patients.[15] And as Sommer has pointed out, if females also use their right hemisphere for language, they should have more language problems after right-hemisphere damage than do men. But they don't.[16]

So are males really more lateralised for language? It's not clear why one would think so. And if men *are* more lateralised, it doesn't seem to do them much harm. Several researchers have recently argued that gender differences in language skills are actually more or less nonexistent.[17]

The supposedly larger female corpus callosum, a claim built on shaky foundations, is under no less serious dispute.[18] This research has been thoroughly examined and critiqued by Brown University professor of biology Anne Fausto-Sterling who, in *Sexing the Body*, explains the challenges of establishing the size of a particular structure in the brain. And a meta-analysis conducted by Katherine Bishop and Douglas Wahlsten in 1997 concluded that 'the widespread belief that women have a larger splenium than men and consequently think differently is untenable.'[19] Summarising this literature in a 2008 review, cognitive neuroscientist Mikkel Wallentin concluded that 'the alleged sex-related corpus callosum size difference is a myth.' The culprit? Look no further than 'the possibility of "discovering" spurious differences when using small sample sizes', says Wallentin.[20]

So let us, with healthy scepticism, summarise all of this as clearly as we can. Nonexistent sex differences in language lateralisation, mediated by nonexistent sex differences in corpus callosum structure, are widely believed to explain nonexistent sex differences in language skills.

Confused?

We've only just begun.

The picture becomes only more puzzling when we look for evidence that men are more lateralised for visuospatial tasks. Some neuroimaging studies have found more lateralised activation in men of the parietal areas thought to be especially involved in this kind of processing. But others find no sex differences, and yet others find more lateralisation of activity in women.[21]

Yet variations on a theme that contrasts a female, 'floodlight' brain that is global and interhemispheric in processing style with a male, 'spotlight' brain that is localised and intrahemispheric are everywhere. For example, a consensus statement titled 'The Science of Sex Differences in Science and Mathematics' links female 'interhemispheric connectivity' to an advantage in language skills and male within-hemisphere connectivity to superiority in 'tasks requiring focal activation of the visual association cortex', that is, visuospatial tasks.[22]

Simon Baron-Cohen has also taken up the spotlight/floodlight dichotomy. He and his colleagues tentatively suggested in an article in *Science* that the male brain skew towards 'increased local connectivity' makes it better suited to understanding and building systems. By contrast, the female brain skew towards 'long-range' and 'interhemispheric connectivity' is better structured for empathising.[23] And Ruben Gur, a professor of psychiatry at the University of Pennsylvania who coined the floodlight/spotlight metaphor, explained to a journalist for the *LA Times* that brain science tells us that '[i]n a stressful, confusing multi-tasking situation, women are more likely to be able to go back and forth between seeing the more logical, analytic, holistic aspects of a situation and seeing the details,' while 'men will be more likely to deal with [the situation] as, "I see/I do, I see/I do, I see/I do."'[24] The implications of this difference for mental juggling may explain why Gur's wife and collaborator, Dr. Raquel Gur, must take on the main burden of quickly putting together a meal for a hungry family. Gur *can* throw together a salad '[b]ut', he says, 'I can't at the same time worry about whether this is in the microwave and that is in the skillet. When I do, something will burn.'[25] Presumably, in that sad pile of cinders also lie the smoldering ashes of Mrs. Gur's hopes of someone else ever being in charge of the meals.

Little surprise, then, with such scientific endorsements, to find popular writers picking up these ideas and running with them. Michael Gurian, whose Gurian Institute offers training to teachers, parents and corporations, becomes impressively quantitative on the topic, explaining to educators that '[b]ecause boys' brains have more cortical areas dedicated to spatial-mechanical functioning, males use, on average, half the brain space that females use for verbal-emotive functioning.'[26] Meanwhile, Allan and Barbara Pease take the lateralisation hypothesis to its natural extreme in their book *Why Men Don't Listen and Women Can't Read Maps*, by claiming that the female brain is *so* unlocalised for spatial processing that it doesn't even *have* 'a specific location for spatial ability'[27] – thus neatly furnishing an answer to the second part of the title of their book. And why stick to language and visuospatial skills when, as certain academics have shown us, any gender stereotype can be pinned to sex differences in hemisphere use, in impressively scientific-sounding fashion? For instance, what began as women's supposedly more bilateral language skills quickly transformed into the basis of womanly intuition and multitasking skills while, as John Gray explains in *Why Mars and Venus Collide*, men's more localised brain activity even explains their propensity to forget to buy milk.[28]

But in all the excitement of having found a neurological explanation for male inconsiderateness and female underrepresentation in the Faculty of Mental Rotation, people failed to notice that the empirical ground had shifted beneath their feet. And they also forgot to ask a very important question: Why should a localised *brain* create a spotlight *mind* good at certain masculine tasks? And why should a global, interconnected *brain* create a floodlight *mind* better at feminine activities?[29] And this brings us to the second problem with interpreting sex differences in the brain: what do they actually mean for differences in the mind?

> Seeing that the average brain-weight of women is about five
> ounces less than that of men, on merely anatomical grounds we
> should be prepared to expect a marked inferiority of intellec-
> tual power in the former. Moreover, as the general physique of
> women is less robust than that of men – and therefore less able
> to sustain the fatigue of serious or prolonged brain action – we
> should also on physiological grounds be prepared to entertain
> a similar anticipation. In actual fact we find that the inferiority
> displays itself most conspicuously in a comparative absence of
> originality, and this more especially in the higher levels of intel-
> lectual work.
>
> —George J. Romanes, evolutionary biologist and physiologist (1887)[1]

It's always pleasant when data confirm predictions. But did George Romanes never once consider whether an African Grey parrot (with a brain weight of less than half an ounce) might out-smart a cow with a brain more than thirty times heavier? Did he really know not a single weedy intellectual, nor one muscular chump, to provoke him to wonder whether physical strength really was correlated with tenacity of 'brain action'? Perhaps it was only natural that the brain scientists who meticulously measured men's and women's head dimensions, skull volume and brain weight should try to relate their findings to psychological differences between the sexes. But with the benefit of hindsight we can see that it was not just neuroscientific understanding they lacked, but humility. 'Optimistic' is the only kind word to use to describe their confident assertions that differences in the engine power of male and female *minds* were being probed by tape measures, sacks of millet grain and sets of scales.

Today, we are no less interested in pinning our more sophisticatedly obtained sex differences in the brain onto the mind. '[H]ope springs eternal', Fausto-Sterling wryly notes. 'Is it now possible that finally, with *really* new, *really* modern approaches, we can demonstrate the biological basis of sexual or racial inequality?'[2] And, as neuroendocrinologist Geert De Vries has pointed out, it is intuitive to assume that males and females have different brains so that they can behave differently. With the discovery of differences in hormone receptors, or neuronal density, or corpus callosum size, or different proportions of grey and white matter, or brain region size, the instinct is to look for a psychological difference to pin it on. But the counterintuitive possibility that always needs to be considered is that sex differences in the brain may also 'just as well do the exact opposite, that is, they may prevent sex differences in overt functions and behavior by compensating for sex differences in physiology.'[3] For example, a smaller number of neurons in a particular brain region can be compensated for by greater neurotransmitter production per neuron.[4]

One very striking example of the principle that brain difference can yield behavioural similarity, discussed by De Vries, comes from the prairie vole. In this species, males and females contribute equally to parenting (excepting, of course, nursing). In female prairie voles, parenting behaviour is primed by the hormonal changes of pregnancy. But this leaves a mystery. How do father voles, which experience none of these hormonal changes, come to show paternal behaviour? The answer turns out to lie in a part of a region of the brain called the lateral septum, which is involved in the triggering of paternal behaviour. This part of the brain is very different in males and females, being much more richly endowed with receptors for the hormone vasopressin in the male, yet this striking sex difference in the brain enables male and female prairie voles to behave the same. We can't assume that even quite sub-stantial sex differences in the brain imply sex differences in the mind. As Celia Moore has pointed out, 'Some neural differences are inconsequential, because they are offset by other compensatory differences. Other neural differences are alternative pathways to the same behavioral end.'[5]

In humans, one indisputable physiological difference between males and females is size – including the brain. Although there is overlap, men on average have larger brains than do women, and a large brain is not simply a smaller brain scaled up. Larger brains create different sorts of engineering problems and so – to minimise energy demands, wiring costs and communication times – there are physical reasons for different arrangements in differently sized brains.[6] From this perspective, 'men and women confront similar cognitive challenges using differently sized neural machinery.'[7] The brain can get to the same outcome in more than one way. And in line with this, recent studies of brain structure have argued that it is not that *women* have larger corpora callosa, or a more generous serving of grey matter, relative to brain volume. Rather, it is people with *small brains*, male or female, who show this quality. As one group put it: 'brain size matters more than sex.'[8] If this principle proves to be correct – there's currently no agreed way of controlling for absolute brain size – then, unless we're happy to start com-paring the spatial or empathising skills of big-headed men and women with those of their pin-headed counterparts, we may have to abandon the idea that we will find the answers to psychological gender differences in grey matter, white matter, corpus callosum size or any other alleged sex difference in brain structure that turns out to have more to do with size than sex.

This, one would think, would secretly be a relief. This is not just because those gender differences can wax and wane, depend-ing on the time, place and context. But also the very idea of try-ing to relate these kinds of structural differences to psychological function is fantastically ambitious, given that, as neuroscientist Jay Giedd and colleagues have put it, 'most brain functions arise from distributed neural networks and that within any given region lies a daunting complexity of connections, neurotransmitter systems, and synaptic functions'.[9]

Yet sometimes the temptation is too much to resist.

Twenty years ago, my mother proposed a neuroscientific model to explain why some brains have an extraordinary capacity for deeply focused thought. Her hypothesis was that '[a]ll the blood in your brain rushes to the really clever bits and there's none left over to warm up the roots.'[10] My mother, by the way, is a novelist. Yet her idea, coined as an acerbic marital insult in a work of fiction, shares an important flaw with a suggestion made in a prestigious journal of science. Simon Baron-Cohen and his colleagues, as mentioned earlier, suggested in *Science* that a brain skewed towards local connectivity is 'compatible with strong systemizing, because systemizing involves a narrow attentional focus to local information, in order to understand each part of a system.'[11] Likewise, in the recent book *Why Aren't More Women in Science?* neuroscientists Ruben and Raquel Gur conjecture that 'the greater facility of women with interhemispheric communications may attract them to disciplines that require integration rather than detailed scrutiny of narrowly characterised processes.'[12]

But why, we might ask, should shorter circuits in the brain allow narrower focus in the mind? As McGill University philosopher of science Ian Gold has said, '[m]ay as well say hairier body so fuzzier thinker. Or that human beings are capable of fixing fuses because the brain uses electricity.'[13] Consider what's involved in zooming in your attention on, say, a small aspect of the process of photosynthesis. Does only a little bit of the brain get involved because only a little detail is being processed? Or is there – as seems far more likely – activity all over the brain as distracting information is suppressed, the inner voice formulates ideas and poses questions, visual stimuli are processed, motion is imagined and information is retrieved from memory?[14]

In truth, if it was the male brain that seemed to be more long-range, we could easily concoct a plausible hypothesis to explain why this enhances their systemising skills. And this is the problem: the obscurity of the relationship between brain structure and psychological function means that just-so stories can be all too easily written and rewritten. Do you find that your male participants are actually *less* lateralised on a spatial problem? Not to worry! As the contradictory data come in, researchers can draw on both the hypothesis that men are better at mental rotation because they use just one hemisphere, as well as the completely contrary hypothesis that men are better at mental rotation because they use *both* hemispheres. So flexible is the theoretical arrangement that researchers can even present these opposing hypotheses, quite without embarrassment, within the very same article.[15]

Likewise, Gur and his colleagues happily tinker with the long-standing idea that it is males' more lateralised spatial processing that underlies their superiority on mental rotation tasks. They found that performance on two spatial tasks correlated with the volume of interconnecting white matter in the brain.[16] White matter is made up of the axons, insulated for speed of travel of the electrical signal by the white fat myelin, which communicate between distant brain regions. 'When we looked at the top performers for spatial tasks in our study . . . there were nine men and only one woman,' Gur explained for the *Science Daily* news release. 'Of these nine men, seven [actually, it was six] had greater white-matter volumes than any of the women in the study.'[17] Now, we're talking about ten people here – hardly a sample size on which to base sweeping generalisations about the sexes. It's also, as psychologists well know, dangerous to assume that correlation means causation. Further, in the scientific article itself, Gur cautions that the 'correlations could be spurious and should be interpreted with extreme caution.'[18] And they really could be spurious, given that 1 in 20 'significant' results occur by chance, and the researchers tested for thirty-six relationships. Of course, we don't know who decided that this caveat was not worth mentioning in the report designed for public consumption. But despite all this, Gur goes on to suggest to *Science Daily* that 'in order to be a super performer in that area, one needs more white matter than exists in most female brains.' Following up this line of argument in their chapter in *Why Aren't More Women in Science?* the Gurs conjecture that '[t]he requirement of large volume of WM [white matter] for complex spatial processing may be an obstacle in some branches of mathematics and physics.'[19] This, they suggest, is because men's greater white matter volumes enable better within-hemisphere processing.

But meanwhile, back in the functional neuroimaging lab, the Gurs and their colleagues have found that in some regions of the brain men show more *bilateral* activation than women while performing spatial tasks. They therefore suggest a 'reformulation' of the spotlight hypothesis, namely, 'that optimal performance requires both unilateral activation in primary regions, left for verbal and right for spatial tasks, and bilateral activation in associated regions.'[20] Well, maybe they are right to now emphasise the importance of participation from both hemispheres. Interestingly, researchers who study people with exceptional talent in mathematics argue that enhanced interaction *between* the hemispheres – supposedly a female brain characteristic – is a special feature of the mathematically gifted brain.[21] But maybe, just until such a time as we have a somewhat firmer grasp of how the structural properties of the brain relate to complex cognition, the Gurs should stick to the lower-maintenance hypothesis that optimal performance requires whatever features of the brain happen to be observed in males.[22]

This kind of theoretical U-turn has always beset the neuroscience of sex differences. For example, in the nineteenth century, when the seat of the intellect was thought to reside in the frontal lobes, careful observation of male and female brains revealed that this region appeared both larger and more complexly structured in males, while the parietal lobes were better developed in women. Yet when scientific thought came to the opinion that it was instead the parietal lobes that furnished powers of abstract intellectual thought, subsequent observations revealed that the parietal lobes were more developed in the male, after all.[23] With startling insight, Havelock Ellis, the author of a comprehensive late-nineteenth-century review of sexual science, described these earlier erroneous observations as 'inevitable':

> It was firmly believed that the frontal region is the seat of all the highest and most abstract intellectual processes, and if on examining a dozen or two brains an anatomist found himself landed in the conclusion that the frontal region is relatively larger in women, the probability is that he would feel he had reached a conclusion that was absurd. It may, indeed, be said that it is only since it has become known that the frontal region of the brain is of greater relative extent in the ape than it is in Man, and has no special connection with the higher intellectual processes, that it has become possible to recognise the fact that that region is relatively more extensive in women.[24]

Of course, there's nothing wrong with changing your mind in the light of new evidence about the sexes. But those who are tempted to play this game, by claiming that sex differences in the structure of the brain yield essentially different kinds of minds, should be aware that this sort of flipping seems to be a common part of the process. And, with the benefit of hindsight, it never looks good.

No less care is required when it comes to interpreting differences between the sexes in brain activity. No doubt about it, functional neuroimaging technologies have brought the fresh, modern zing of neuroscience to old stereotypes. Allan and Barbara Pease, for example, purport to demonstrate in their book *Why Men Don't Listen and Women Can't Read Maps* the striking sex differences in the sheer volume of brain devoted to emotion processing. A brain diagram of 'Emotion in men' shows two blobs in the right hemisphere. As the text explains, emotion in men is highly compartmentalised, meaning that 'a man can argue logic and words (left brain) and then switch to spatial solutions (right front brain) without becoming emotional about the issue. It's as if emotion is in a little room of its own'. But in the illustration of

'Emotion in women' there are more than a dozen blobs scattered across both hemispheres of the brain. What this means, according to the Peases, is that 'women's emotions can switch on simultaneously with most other brain functions'. Or, to call a spade a spade, emotion can cloud all and any of a woman's mental activities.[25]

These emotion maps of the male and female brain, the Peases inform readers, are based on fMRI research by neuroscientist Sandra Witelson. In order 'to locate the position of emotion in the brain', she used 'emotionally-charged images that were shown first to the right hemisphere via the left eye and ear and then to the left hemisphere via the right eye and ear.'[26] Should readers have both the time and the resources to check out the six Witelson references in the book's bibliography, they will find only two studies published after functional neuroimaging techniques first began to be substantively put to use by cognitive neuroscientists in the 1980s. One study did not involve brain research (it is a survey of handedness in gay men and women). The other is a comparison of corpus callosum size in right- and mixed-handed people.[27] It might also be worth mentioning that it was a postmortem study. Possibly Sandra Witelson really did present her samples of dead brain tissue with emotionally charged images – but if she did, it's not mentioned in the published report.

It may be that the Peases were referring to functional neuroimaging research published by Sandra Witelson and colleagues in 2004.[28] It's hard to know: this study used PET rather than fMRI; stimuli were presented in the normal two-eyed, two-eared fashion; and the male/female blob tallies and locations are dissimilar to those presented by the Peases. However, this study did at least look at brain activity while men and women performed one of two emotion-matching tasks. The easier task involved deciding which of two faces match the emotion of a third, target, face. The harder task involved deciding which of two faces match the emotion expressed in a voice. According to Susan Pinker's summary of Witelson's results, '[w]hen women looked at pictures of people's facial expressions, both cerebral hemispheres were activated and there was greater activity in the amygdala, the almond-shaped seat of emotion buried deep in the brain. In men, perception of emotion was usually localised in one hemisphere'. Pinker then goes on to suggest that since research also shows that women have a thicker corpus callosum, allowing speedy interhemispheric transmission of information (a claim that, as you will recall from the previous chapter, is under serious scientific dispute), 'the hardware for women's processing of emotion seems to take up more space and have a more efficient transportation grid than men's. Scientists infer that this allows women to process emotion with dispatch.'[29]

In fact, the researchers found no differences in how quickly men and women performed the tasks. It's also worth noting that although the statement 'both cerebral hemispheres were activated' in women might conjure up an image much like that presented by the Peases, with activity over a generous portion of the female brain, this is not the case. Rather – and take a deep breath before reading on – in the easy task women showed greater activation than men in left fusiform gyrus, right amygdala and left inferior frontal gyrus. In the hard task they showed greater activity in left thalamus, right fusiform gyrus and left anterior cingulate. Men, meanwhile, showed greater activity than women in right medial frontal gyrus and right superior occipital gyrus for the easy task, and in left inferior frontal gyrus and left inferior parietal gyrus for the hard task. Or, rather less technically, women always had two left blobs and one right blob, while men had either two right blobs or two left blobs, depending on the task – painting a rather less striking image of contrast. (Bear in mind, too, that blobs represent *differences* in brain activity, not brain activity per se. If a search for regions activated more in men yields a blob-free left hemisphere, for example, that doesn't mean that that hemisphere is switched off in men. Rather, it means that the researchers didn't find any regions in the hemisphere that were activated more in men than in women.)[30]

Does this complicated-sounding list of brain activations tell us something interesting about gender difference in emotional experience? The researchers, like Pinker, certainly think so. They conclude that their 'findings suggest that men tend to modulate their reaction to stimuli, and engage in analysis and association, whereas women tend to draw more on primary emotional reference.'[31] (By this they mean that only women find others' emotions innately arousing.) As you will have already realised, a simpler, and more familiar, way to put the same idea would be to say that men are thinkers and women are feelers.

So does this neuroimaging study simply confirm what everyone already suspected – that 'men may take a more analytic approach' to emotion processing while 'women are more emotionally centred'?[32] Or is it possible that these interpretations are, to paraphrase Fausto-Sterling, unwittingly projecting assumptions about gender onto the vast unknown that is the brain?

With the previous chapter's cautionary tale of premature speculation in mind, it's worth noting that Witelson's neuroimaging study compared just eight men with eight women on each task – a modest-sized sample. Could the sex differences in brain activation be spurious? When looking for changes in blood flow between two conditions, researchers search in thousands of tiny sections of the brain (called voxels), and many researchers are now arguing that the threshold commonly set for declaring that a difference is 'significant' just isn't high enough. To illustrate this point, some researchers recently scanned an Atlantic salmon while showing it emotionally charged photographs. The salmon – which, by the way, 'was not alive at the time of scanning' – was 'asked to determine what emotion the individual in the photo must have been experiencing.' Using standard statistical procedures, they found significant brain activity in one small region of the dead fish's brain while it performed the empathising task, compared with brain activity during 'rest'. The researchers conclude not that this particular region of the brain is involved in postmortem piscine empathising, but that the kind of statistical thresholds commonly used in neuroimaging studies (including Witelson's emotion-matching study) are inadequate because they allow too many spurious results through the net.[33]

This of course does not mean that all reported activations are spurious. It just highlights the importance of being aware of the possibility. We might be more confident that Witelson's study genuinely identified brain regions that function differently in the two sexes during emotion recognition tasks if at least some of the brain regions that showed sex differences in activation in the easy emotion-matching tasks also turned up in the harder task.[34] However, if you look back at the list of brain activations you'll see that in neither men nor women was any brain region activated more during both the easy and difficult emotion-matching tasks.

But even if we assume that results such as these are reliable, what do they tell us about male/female differences in psychology? Does it mean that men are more analytic, if their left inferior frontal gyrus activates more, or that women are more emotional because the right amygdala is on fire? Inferring a psychological state from brain activity (like *The amygdala was activated so that means our participants were fearful*) is known as reverse inference, and as any neuroimager will tell you, it is fraught with peril.[35] Some neuroscientists have even died while making reverse inferences. Actually, I made that last bit up, but as we will see, it is extremely tricky. There are two ways that males and females can diverge in brain activation: how much activation is seen and where that activation is. Neither piece of information, unfortunately, tells us much about psychological sex differences.

Just as bigger doesn't necessarily mean better with regards to the size of brain structures, neither does more activation necessarily mean better or psychologically more. Researchers who study development, or learning, sometimes find that some patterns of activation reduce, or become more streamlined, as development or expertise proceeds.[36] Bizarrely, activation isn't even a surefire sign that the activity is doing *anything* useful. For example, Chris Bird and colleagues studied a patient who suffered extensive damage to the medial prefrontal cortex following a stroke. The scope of the damage included pretty much all of the brain regions that have been reliably activated in literally dozens of functional imaging studies of mind reading. Yet the patient was fine at mind reading! As the researchers note, 'the data reported here urge caution in concluding that medial frontal cortex is critical for effecting ToM [theory of mind]'.[37] Vision scientist Giedrius Buracas and colleagues had an equally surprising finding. They found that brain region V1 was activated more than region MT in a motion perception task. Yet it's well-established from neurophysiological research with primates that MT – which was activated less – is critically involved in motion detection, while V1 – which was activated more – is not.[38] These two studies serve as warning flags: even though a part of the brain might light up during a task, it may not be especially or crucially involved.

The location of activation in the brain is also surprisingly uninformative. Clearly, the whole brain isn't involved in doing everything. Different parts of the brain are specialised for processing different sorts of information. But a particular cortical region or population of neurons can be specialised for different jobs in different contexts. As imaging experts Karl Friston and Cathy Price put it, specialisation is dynamic and context-dependent.[39] For example, a particular population of neurons in the temporal cortex may, at different times, represent both identity (*Whose face is it?*) and expression (*Is it happy or sad?*). What those neurons are doing depends both on what sort of information is being fed in, and also what sort of information is being fed back from higher regions in the processing chain. 'Specialisation is therefore not an intrinsic property of any region', argue Price and Friston, and that means that seeing a brain region in action doesn't mean you know what it's up to in your particular task. For many parts of the brain, this problem is acute. For example, the anterior cingulate is activated by so many tasks that one cognitive neuroscientist I know refers to it as the 'on button'.

There just isn't a simple one-to-one correspondence between brain regions and mental processes, which can make interpreting imaging data a difficult task. As Jonah Lehrer recently explained in the *Boston Globe*:

[O]ne of the most common uses of brain scanners – taking a complex psychological phenomenon and pinning it to a particular bit of cortex – is now being criticized as a potentially serious oversimplification of how the brain works. . . . [C]ritics stress the interconnectivity of the brain, noting that virtually every thought and feeling emerges from the crosstalk of different areas spread across the cortex.[40]

If so, the familiar spots of colour on brain activation maps (derided by some as 'blobology'), labelled as male-female difference in activation, are going to tell a very oversimplified story, and one in which much of the important information may be lost. It's also a story that, as neuropsychologist Anelis Kaiser and colleagues point out, is geared to emphasise difference over similarity.[41]

Then, there is the sad fact that, at its most precise, functional imaging technology averages over a few seconds the activity of literally millions of neurons that can fire up to a hundred impulses a second. (For PET the time-scale is even longer.) 'Using fMRI to spy on neurons is something like using Cold War–era satellites to spy on people: Only large-scale activity is visible', says *Science* journalist Greg Miller.[42] This severely limits the interpretations that can be made about brief psychological events. Understandably, given all these interpretative gaps, many neuroscientists hesitate to speculate what their data might mean in terms of sex differences in thinking. Many, to their credit, have performed admirably as The Voice of Restraint in popular articles about gender and the brain, and in their academic work explicitly warn against making unwarranted inferences (pleas that, in certain quarters, fall on deaf ears).

It's not, by the way, my intention to present myself as a neuroscience sceptic. Not only are some of my best friends, as well as family members, neuroimagers, but I also think that neuroscience is an extremely exciting and promising field, and can be usefully employed in combination with other techniques. I also understand that speculation is an important part of the scientific process. Nor is the topic of gender difference by any means the only area in which overinterpretation can occur. And I certainly don't think that research into sex differences in the brain is wrong or pointless. There *are* sex differences in the brain (although, as we've seen, agreeing on what these are is harder than you might think);[43] there are sex differences in vulnerabilities to certain psychological disorders, and hopefully greater understanding of the former might help to illuminate the latter. My point is simply this: that neither structural nor functional imaging can currently tell us much about differences between male and female minds. As Rutgers University psychologist Deena Skolnick Weisberg has recently argued, we should 'remember that neuroscience, as a method for studying the mind, is still in its infancy. It shows much promise to be someday what many people want to make it into now: a powerful tool for diagnosis and research. We should remember that it has this promise, and give it the time it needs to achieve its potential – without making too much of it in the meantime.'[44]

Are early twenty-first-century neuroscientific explanations of inequality – too little white matter, an unspecialised brain, too rapacious a corpus callosum – doomed to join the same garbage heap as measures of snout elongation, cephalic index and brain fibre delicacy? Will future generations look back on early twenty-first-century interpretations of imaging data with the same shocked amusement with which we regard early twentieth-century speculations about the relevance of sex differences in spinal cord size? I suspect they will, although only time will tell. But to any scientist considering trying to relate sex differences in the brain to complex psychological functions . . . well, let's just say, 'Remember Dr. Charles Dana'.

And it *is* important to remember him. For as we'll see in the next chapter, the speculations of a few scientists quickly evolve into the colourful fabrications of popular neurosexism – the subspecialty within the larger discipline of neurononsense to which we now turn.

My husband would probably like you to know that, for the sake of my research for this chapter, he has had to put up with an awful lot of contemptuous snorting. For several weeks, our normally quiet hour of reading in bed before lights out became more like dinnertime in the pigsty as I worked my way through popular books about gender difference. As the result of my research, I have come up with four basic pieces of advice for anyone considering incorporating neuroscientific findings into a popular book or article about gender: (1) unless you have a time machine and have visited a future in which neuroscientists can make reverse inferences without the nagging anxieties that keep the more thoughtful of them awake at night, do not suggest that parents or teachers treat boys and girls differently because of differences observed in their brains; (2) if you don't know what a reverse inference is, read the previous chapter of this book; (3) exercise extreme caution when making the perilous leap from brain structure to psychological function; and (4) don't make stuff up.

When it comes to selecting examples from those who have failed to follow one or more of these four simple rules, one's choices abound. Possibly my favourite illustration of a self-serving projection of prejudices onto brain jargon is a section in John Gray's *Why Mars and Venus Collide* in which he discusses the inferior parietal lobe (IPL). In men, says Gray, the left IPL is more developed, while in women it is the right side that is larger. It will be no surprise to anyone, I am sure, to learn that '[t]he left side of the brain has more to do with more linear, reasonable, and rational thought, while the right side of the brain is more emotional, feeling, and intuitive.' But it *is* extraordinary just how differently the IPL serves its master and its mistress. According to Gray a man's large left IPL, being involved in the 'perception of time', explains why he becomes impatient with how long a woman talks. By contrast, the IPL also 'allows the brain to process information from the senses, particularly in selective attention, like when women are able to respond to a baby's crying in the night.'[1] Perhaps deliberately, we are left in the dark as to whether the male inferior parietal lobe enables a man to do the same.

In *Leadership and the Sexes*, Michael Gurian and Barbara Annis inform executives that 'women's brains tend to link more of the emotional activity that is going on in the middle of the brain (the limbic system) with thoughts and words in the top of the brain (the cerebral cortex). Thus a man might need many hours to process a major emotion-laden experience [*I . . . just . . . got . . . fired. . . . I . . . am . . . sad . . . and . . . angry.*], whereas a woman may be able to process it quite quickly [*Oh, crap!*].'[2] A further neurophysiological disadvantage for men may be found in another of Gurian's books, *What Could He Be Thinking?* Implicitly drawing on a working metaphor of *The Brain as Pinball Machine*, he explains how in men the 'signal' of an emotional feeling, having made it to the right hemisphere, 'may well get stopped, disappearing into neural oblivion because the signal found no access to a receptor in a language center in the left side of the brain.' This doesn't happen in the female brain because, according to Gurian, while men have just one or two language centres in the left hemisphere, women have as many as seven such centres, dotted all over the brain, as well as a 25 percent larger corpus callosum. (Despite this embarrassment of neurological riches, the contrast Gurian draws between male and female brain function leaves me speechless.) And so, in men, a feeling signal is much less likely to hit the jackpot of contact with a neuron involved in language.[3]

We also discover in *Leadership and the Sexes* that when a woman

leader asks her colleagues, 'What do you all think?' this is a typically female 'white matter' question. It seems that white matter
isn't just involved in integrating information from different parts
of the brain, but also from different people in the office.[4] Brain differences may also be behind a female-leadership problem-solving
style: when a female leader 'knows what to do, she's not as worried
as a man might be about proving it with data'. Gurian and Annis
suggest that '[o]ne reason for this intuitiveness may be that she has
a larger *corpus callosum* connecting both hemispheres of the brain'.
By contrast, male leaders favour a problem-solving style that, in
part, 'relies on more linear data and proof.'[5]

Perhaps my own corpus callosum runs to a smaller size than
the standard female issue, but I find these intuitive leaps from brain
structure to psychological function unconvincing, as noted in the
previous chapter. Why should arriving at a solution to a problem
through an analysis of data and proof require any less integration
between hemispheres? As an example of just how wrong our intu-
itions can be in these matters, despite the popular assumption that
a more lateralised brain will be worse at multitasking, neurobiologist Lesley Rogers and her colleagues found precisely the opposite
to be the case in chicks.[6] Chicks with more lateralised brains were
better at simultaneously pecking for food grains and looking out
for predators (the established chick equivalent of frying a steak
while making a salad).

While it may not be too surprising to discover self-appointed
'thought-leaders' dressing up stereotypes in neuroscientific finery, it is more of a shock to see this in an alumnus of Harvard
Medical School, the University of California–Berkeley, and Yale
School of Medicine. Step forward Louann Brizendine, director of
the University of California–San Francisco Women's Mood and
Hormone Clinic. Her book, *The Female Brain*, cites literally hun-
dreds of academic articles. To the unwary reader, both she and the
book seem reliable and authoritative. And yet, as a review of the
book in *Nature* comments, 'despite the author's extensive academic
credentials, *The Female Brain* disappointingly fails to meet even

the most basic standards of scientific accuracy and balance. The book is riddled with scientific errors and is misleading about the processes of brain development, the neuroendocrine system, and the nature of sex differences in general.' The reviewers later go on to say that, '[t]he text is rife with "facts" that do not exist in the supporting references.'[7] This is a common discovery made by people who take the time to fact-check Brizendine's claims. Mark Liberman, a professor at the University of Pennsylvania with no special interest in gender issues, has nonetheless been provoked to provide many detailed but humorous critiques of pseudoscientific claims about gender differences on his online Language Log. His patient corrections of Brizendine's many false assertions about sex differences in communication is a chore that, as he puts it, 'is starting to make me feel like the circus clown that follows the elephant around the ring with a shovel.'[8]

But despite these forewarnings, when I decided to follow up Brizendine's claim that the female brain is wired to empathise, it nonetheless proved to be an exercise that turned up surprise after surprise. I tracked down every neuroscience study cited by Brizendine as evidence for feminine superiority in mind reading. (No, really, no need to thank me. I do this sort of thing for pleasure.) There were many such references, over just a few pages of text, creating the impression it was no mere opinion, but scientifically established fact, that the female brain is wired for empathy in a way that the male brain is not. Yet fact-checking revealed the deployment of some rather misleading practices. For example, let's work our way through the middle of page 162 to the top of page 164 in her book. We kick off with a study of psychotherapists, which found that therapists develop a good rapport with their clients by mirroring their actions.[9] Casually, Brizendine notes, 'All of the therapists who showed these responses happened to be women.'[10] For some reason, she fails to mention that this is because only female therapists, selected from phone directories, happened to be recruited for the study.

Brizendine's next claim – that girls have an advantage in

understanding others' feelings – does find support in the work of Erin McClure and Judith Hall, which she cites. These researchers both conducted meta-analyses that found advantages for females in decoding nonverbal expressions of emotion.[11] The edge is, however, moderate. McClure's meta-analysis suggests that about 54 percent of girls will perform above average in facial emotion processing, compared with 46 percent of boys. Hall's review of research with tests such as the PONS nonverbal decoding task (which we encountered in Chapter 2) suggests that if you randomly chose a boy and a girl, over and over, more than a third of the time the boy would outperform the girl. Brizendine does not understate these findings, then, when she says that '[g]irls are years ahead of boys' in these abilities.[12] She then speculates that mirror neurons may lie behind these skills, enabling girls to observe, imitate and mirror the nonverbal cues of others as a way to intuit their feelings. (Mirror neurons are neurons that respond to another animal's actions as though the animal-observer itself were acting. Some scientists think that mirror neurons may provide the neural grounding for understanding people's minds. Other scientists are dubious about the whole concept.) The study she cites here does explore the potential role of the mirror system in intuiting others' mental states – but not specifically in females.[13] Indeed, its participants (some of whom had autism-spectrum disorders) were all male.

A little later, readers are told that 'brain-imaging studies show that the mere act of observing or imagining another person in a particular emotional state can automatically activate similar brain patterns in the observer – and females are especially good at this kind of emotional mirroring.'[14] Cited as support for this feminine superiority in emotional mirroring is a 2004 neuroimaging study by cognitive neuroscientist Tania Singer and colleagues, who compared brain activation when someone was either receiving a painful electric shock to the hand or was aware that a loved one was receiving the same painful electric shock to the hand.[15] Singer and colleagues found that some brain regions were activated both by being shocked and watching someone else be shocked. If you think I'm going to be nitpicky about what any sex differences in activation in this study *mean*, you're wrong. Actually, the problem of interpretation is rather more basic. Only women were scanned.

Continuing the theme of women's special sensitivity to the pain of others in the next paragraph, Brizendine informs us that when a woman, for example, responds empathically to the stubbed toe of another, she is 'demonstrating an extreme form of what the female brain does naturally from childhood and even more in adulthood – experience the pain of another person.'[16] Brizendine marshals two functional neuroimaging studies as support for this claim. The first is Singer's 2004 study of females' empathic responses to pain. The second is a study by Tetsuya Iidaka and colleagues, who asked participants to judge the gender of faces showing positive, negative or neutral expressions. They compared brain activations in young versus old participants, but not in females versus males.[17] (Her third citation is a review of anxiety and depression in childhood and adolescence. It doesn't discuss responses to others' pain, or gender differences in this capacity, although the authors note that '[b]ecause females are known to be more emotionally responsive than males to the problems of *others*, a wider range of interpersonal contexts may arouse them.')[18]

In the last part of this page range, Brizendine describes Singer's 2004 study, and states that 'the same pain areas of [the women's] brains that had activated when they themselves were shocked lit up when they learned their partners were being strongly shocked.'[19] She references the Singer 2004 study here, naturally, but also another functional neuroimaging study by the same research team, published in 2006.[20] This study was similar, but instead of being a romantic partner who was shocked, it was a confederate who had played either fairly or unfairly in a game just before. In this study, both men and women were scanned. Again, empathy-related responses were seen in reaction to the pain of another, although in men this was only the case when the confederate had played fairly. Having referenced these two studies, Brizendine concludes that '[t]he women were feeling their partner's pain. . . . Researchers have been unable to elicit similar brain responses from men.'[21] She has, however, just cited a study that *did* elicit similar brain responses from men, albeit only in response to people they liked.

By this point the reader may have a poor opinion indeed of the male neurological capacity for empathy – especially since earlier on in the chapter Brizendine suggests that females may have more of the neurons that enable mirroring. She writes that '[a]lthough most of the studies on this topic have been done on primates, scientists speculate that there may be more mirror neurons in the human female brain than in the human male brain.' Look to the notes at the back of the book and no fewer than five scholarly references appear to affirm this claim.[22] The first study is in Russian. Although it did compare the sexes, from the abstract I would lay a substantial bet on it not offering much insight into gender differences in mirror neurons, as it was a postmortem study of neuron characteristics in the frontal lobes. (One would, I imagine, have to see mirror neurons in action to be able to identify them.) Three further studies did indeed look at some aspect of what is thought to be the mirror neuron system. However, none of them compared males and females, or speculated about possible differences between the sexes. And that leaves just one remaining citation, which is 'personal communication' with Harvard-based cognitive neuroscientist Lindsay Oberman, entitled 'There may be a difference in male and female mirror neuron functioning'. When I emailed Dr. Oberman to confirm, to my surprise, she informed me that not only had she never communicated with Brizendine, but went on to write that, 'to the contrary, I have looked at many of my studies and have not found evidence for better mirror neuron functioning in females.'[23] (Once you've picked your jaw up off the floor, don't forget to briefly think about the 5 percent rule I mentioned in Chapter 12, in which only sex *differences* get reported.)

What is deliciously ironic about all of this is that Brizendine presents herself as the reluctant but fearless messenger of truth:

> In writing this book I have struggled with two voices in my head – one is the scientific truth, the other is political correctness. I have chosen to emphasize scientific truth over political correctness even though scientific truths may not always be welcome.[24]

When I am in the mood to be irked, I flip through Brizendine's book. Perhaps because of the particular stage of life I happen to be in, I found myself most enraged by her claim that only when 'the children leave home, the mommy brain circuits are finally free to be applied to new ambitions, new thoughts, new ideas.'[25] But it's the sexism that bursts through the doors of preschools and schools, cleverly disguised in neuroscientific finery, that I find most disturbing. As neuroimaging takes its first steps on the long journey to understanding how neuronal firing yields mental abilities, you will find no shortage of so-called experts willing to explain the educational implications of differences in boy wiring and girl wiring. The medal for the most outrageous claim must surely go to an American educational speaker. According to reports sent to Mark Liberman's Language Log, this educational consultant has been informing audiences that girls see the details while boys see the big picture because the 'crockus' – a region of the brain that does not exist – is four times larger in girls than in boys.[26]

I should reassure you that most people who talk about the educational implications of sex differences in the brain do limit themselves to regions recognised by the majority of the scientific community. I also have little doubt that many of them have the very best intentions behind their use of the brain science literature. They want to improve educational outcomes for children of both sexes. Those who promote single-sex schools may certainly have good reasons for their cause that have nothing to do with the brain. But promoting that cause by projecting gender stereotypes onto brain data is worse than useless.

Perhaps the most influential of this group of educational speakers is Leonard Sax of the National Association for Single Sex Public Education (NASSPE), and author of two books that argue a brain-based need for single-sex schooling. Sax has a punishing speaking schedule, that so far has included the United States, Canada, Australia and New Zealand, as well as countries in Europe – and some schools are clearly impressed. NASSPE has been involved in about half of the 360 single-sex public school programmes in the United States, and Sax has told *New York Times* journalist Elizabeth Weil that about 300 of them 'are coming at this from a neuroscience basis.'[27] Let's take a closer look at what that means.

Take an English class, for example. In the girls' class, you will find teachers asking their students to reflect on story protagonists' feelings and motives: *how would you feel if? . . .* sort of questions. But not in the boys' classroom, because '[t]hat question requires boys to link *emotional* information in the amygdala with *language* information in the cerebral cortex. It's like trying to recite poetry and juggle bowling pins at the same time. You have to use two different parts of the brain that don't normally work together.' The problem for boys and young children, according to Sax, is that emotion is processed in the amygdala, a primitive, basic part of the brain – 'that makes few direct connections with the cerebral cortex.'[28] (In fact, the amygdala appears to be richly interconnected with the cerebral cortex.)[29] This supposedly renders them incapable of talking about their feelings. But in older girls, emotion is processed in the cerebral cortex, which conveniently enables them to employ language to communicate what they're feeling. The implications for teaching are clear: *girls to the left, phylogenetically primitive ape-brains to the right!* Yet this 'fact' about male brains – variants of which I have seen repeated several times in popular media – is based on a small functional neuroimaging study in which children stared passively at fearful faces.[30]

It's doubtful whether any negative emotion was involved during the study (except perhaps boredom);[31] the children were not asked to speak or talk about what they were feeling and, critically, brain activity was not even measured in most of the areas of the brain involved in processing emotion and language.[32] As Mark Liberman has pointed out, 'the disproportion between the reported facts and Sax's interpretation is spectacular.'[33] Even if studies *did* show what Sax claims (questionable),[34] why on earth would we assume that the language parts of the brain wouldn't get involved if the child wished to speak? Shifting information from A to B is, after all, what axons and dendrites are *for*. Yet Sax describes with admiration a boy-brain-friendly English class in which boys study *The Lord of the Flies* by reading the text not with an eye on the plot, or characterisation, but so as to be able to construct a map of the island.

And it's all happening at a school near you. At a coeducational school in my neighbouring suburb, 'parallel education' is provided for boys and girls in certain years. As a journalist explains, 'teaching boys [maths] was more about hands-on practice: drawing, doing the exercise. But in a class with girls, Davey [the middle school principal] discusses the issues for a full 10 minutes at the start of the class, while the graph is put in the context of a relationship between two people.'[35] Perhaps Davey has read one of the other 'neurofallacies' propagated by Sax, that because boys process maths in the hippocampus (another one of those primitive parts of the brain that males so seem to favour), but girls process geometry and maths 'in the cerebral cortex' (a statement so unspecific as to be a bit like saying, 'I'll meet you for coffee in the Northern Hemisphere'), this indicates a need for very different educational strategies. Sax claims that because the primitive hippocampus has 'no direct connections to the cerebral cortex' [um, again, not quite right] boys are happy dealing with maths '"for its own sake" at a much earlier age than girls are.' But for the girls, because they're using their cerebral cortex, 'you need to tie the math into other higher cognitive functions.'[36] The goal of inspiring children to get excited about maths is certainly admirable. But Sax's claim that the results of a neuroimaging study of maze navigation point to a brain-based need to teach girls and boys in these different kinds of ways is simply neurononsense.[37]

Mark Liberman has analysed in meticulous detail many of Sax's dubious brain-based educational claims, and has described the way so-called educational experts like Sax and Gurian use scientific data as 'shockingly careless, tendentious and even dishonest. Their over-interpretation and mis-interpretation of scientific research is so extreme that it becomes a form of fabrication.'[38] While it might be amusing to think up romance stories involving stolid Mr. X-Axis and flighty Ms. Y to amuse the girls, or an interesting challenge to discuss a book without mentioning mental states, the danger is that self-fulfilling prophecies are being delivered alongside the new-look, single-sex curriculum.

Vicky Tuck, while president of the Girls' School Association, recently argued that there are 'neurological differences' between the sexes that are 'pronounced in adolescence'. The practical implication? 'You have to teach girls differently to how you teach boys.'[39] Is she right? Remember how easily spurious findings of sex differences can lead to premature speculation. Remember what Celia Moore and Geert De Vries have pointed out – sex differences in the brain can be compensation, or a different path to the same destination. Bear in mind that neuroscientists are still quarrelling over the appropriate statistical analysis of highly complex data. Recall that many sex differences in the brain may have more to do with brain size than sex per se. Remember that psychology and neuroscience – and the way their findings are reported – are geared towards finding difference, not similarity. Male and female brains are of course far more similar than they are different. Not only is there generally great overlap in 'male' and 'female' patterns, but also, the male brain is like nothing in the world so much as a female brain. Neuroscientists can't even tell them apart at the individual level. So why focus on difference? If we focused on similarity, we'd conclude that boys and girls should be taught the same way.

You're not convinced? You feel sure these brain differences must be educationally important? Okay, fine. Separate your boys and girls. Or, if you want to be really thorough, because there is overlap with these sex differences, strictly speaking one should provide separate streaming for, say, Large Amygdalas and Small Amygdalas, or Overactivated versus Underactivated Left Frontal Lobes. And now tell me *how* you tailor your teaching to the size of the amygdala, or to patterns of brain activity to a photo of a fearful face. There is no reliable way to translate these brain differences into educational strategies. It is, as philosopher John Bruer has poetically put it, 'a bridge too far': 'Currently, we do not know enough about brain development and neural function to link that understanding directly, in any meaningful, defensible way to instruction and educational practice. We may never know enough to be able to do that.'[40] And so, instead, we quickly find ourselves falling back on god-awful gender stereotypes.

We never seem to learn.

No discussion of the brain, sex and education would be complete without mention of the now-notorious theory of Professor Edward Clarke of the Harvard Medical School. In his highly successful nineteenth-century book, *Sex in Education* (subtitled, somewhat ironically as it turned out, *Or, A Fair Chance for Girls*), he proposed that intellectual labour sent energy rushing dangerously from ovaries to brain, endangering fertility as well as causing other severe medical ailments.[41] As biologist Richard Lewontin dryly remarked of this hypothesis, 'Testicles, apparently, had their own sources of energy.'[42] From our modern vantage point we can laugh at the prejudice that gave rise to this hypothesis. Yet we may have little cause for complacency.

Tuck says she has 'a hunch that in 50 years' time, maybe only 25, people will be doubled up with laughter when they watch documentaries about the history of education and discover people once thought it was a good idea to educate adolescent boys and girls together.'[43] But when I survey the popular literature, I suspect that this will not be where the people of the future will find their biggest laughs. Frankly, I think they will be too busy giggling in astonished outrage at the claims of early twenty-first-century commentators who, like their nineteenth-century predecessors, reinforced gender stereotypes with crude comparisons of male or female brains; or who, like Brizendine with her talk of 'overloaded brain circuits', attempted to locate social pressures in the brain. (*Here it is, Michael! I finally found the neural circuits for organising child care, planning the evening meal and ensuring that everyone has clean underwear. See how they crowd out these circuits for career, ambition and original thought?*)

I end with a plea. Although, as we'll see in the next chapter, there is something captivating about neuroscientific information, please, no more neurosexism! Follow the four simple steps I set out at the beginning of the chapter or leave the interpretations to the trained professionals. Neuroscience can be dangerous when mishandled, so if you're not sure, be safe.

As the blogger known as Neurosceptic wisely advises those who peddle neurononsense, 'Save yourself . . . put the brain down and walk away.'[44]

I once bought a toy drum that promised to stimulate my child's auditory nerve. I took this to mean that it made noise. Clearly, the genius minds behind the marketing had stumbled on the discovery that information sounds far more impressive when couched in the grand language of neuroscience. (By the way, have I mentioned yet that these words of mine you're reading are stimulating your occipital lobe, as well as refining the neural circuitry of your anterior cingulate gyrus and dorsolateral prefrontal cortex? This isn't just a book – it's a neurological workout.) There's something special about neuroscientific information. It sounds so unassailable, so very . . . well, *scientific*, that we privilege it over boring, old-fashioned behavioural evidence. It brings a satisfying feel to empty scientific explanations. And it seems to tell us who we really are.

After Lawrence Summers's controversial suggestion that women might be inherently less capable of high-level science, Steven Pinker and Simon Baron-Cohen were not the only ones to talk brain differences. So did Leonard Sax. Refreshingly, Sax did not argue that brain research hints at an innate female inferiority, on average, in science and maths. Instead, he argued that the problem lies in an educational system that teaches boys and girls the same things at the same time. This is a mistake because, as he explained in the *Los Angeles Times*, 'while the areas of the brain involved in language and fine motor skills (such as handwriting) mature about six years earlier in girls, the areas involved in math and geometry mature about four years earlier in boys.'[1] Sax argues that teaching should be sensitive to sex differences in the timing of development of the various regions of the brain because '[a] curriculum which ignores those differences will produce boys who can't write and girls who think they're "dumb at math."'[2]

Now, I'm all behind Sax's goal of improving educational outcomes for boys and girls. There might be good reasons for single-sex schooling. But what are we to make of his claim that, as he put it to *CBS News*, '[b]oth boys and girls are being shortchanged as a result of the neglect of hard-wired gender differences'?[3]

By now, you will probably be uneasy about the idea that complex psychological skills like language, maths and geometry can be pinpointed to a single part of the brain. It's simply not the case that people use one particular lobe, or a circumscribed area of the brain, to read a novel, or write an essay, or solve an equation or calculate the angle of a triangle. And, unfortunately, neuroscience has yet to reach the stage at which it can peer into the brain and determine capacity for solving simultaneous equations or readiness to learn calculus. I can understand why this relatively subtle point didn't set off alarm bells in Sax or the editors or journalists who brought comments like this to the public eye. But why did no one query the relevance of Sax's statement on the grounds that boys are clearly *not*, in fact, four years ahead of girls in maths – they are not ahead of them at all, as it happens.[4] Nor, of course, is the language ability of a twelve-year-old boy comparable to that of a six-year-old girl. Even if we are happy to relate one part of the brain to complex cognition, clearly, this concept of neural maturation is a very poor index of actual ability – a far worse measure than, say, a maths test. So why does this kind of neurononsense get column inches?

One reason may be that neuroscience easily outranks psychology in the implicit hierarchy of 'scientificness'.[5] Neuroscience, after all, involves expensive, complex machinery. It generates smart-looking three-dimensional images of the brain. The technicians almost certainly wear white coats. It involves quantum mechanics, for goodness' sake! I ask you, what kind of a match for this is a simple piece of paper on which a six-year-old girl has successfully added 7 and 9? Bioethicist Eric Racine and colleagues coined the term 'neuro-realism' to describe how fMRI coverage can make psychological phenomena somehow seem more real or objective than evidence collected in a more ordinary fashion. They describe how, for example, brain activation in the reward centres of the brain while people ate unhealthy food was provided as evidence that '[f]at really does bring pleasure.'[6] If patterns of firing in the brain can be seen as better proof of someone feeling pleasure than them selecting the box on the questionnaire marked 'Yes, I really enjoyed eating that doughnut', then it's not surprising that children's actual academic skills can be so easily overlooked when brain research is enjoying the spotlight.

I also suspect that because the brain is such a biological organ, with its axons and fat and neurochemicals and electrical impulses, there is the temptation to chalk up whatever sex differences we see in the brain to differences in male and female nature, as Michael Gurian and Kathy Stevens do in *The Minds of Boys*:

> The social thinkers of the 1950s, 1960s, and 1970s did not have PET scans, MRIs, SPECT scans, and other biological research tools available to them. . . . Because they could not look inside the heads of human beings to see the differences in the brains of males and females, they had to lean away from nature-based theory towards social trends theory. They had to overemphasize the power of nurture in gender studies because they didn't have a way to study the actual nature of male and female.[7]

Gurian and Stevens seem to equate 'actual nature' with 'brain'. But really, when you think about it, where else but in the brain would we see the effects of socialisation or experience? As Mark Liberman puts it, 'how else would socially constructed cognitive differences manifest themselves? In flows of pure spiritual energy, with no effect on neuronal activity, cerebral blood flow, and functional brain imaging techniques?'[8] The 'neuro-curmudgeons' from the James S. McDonnell Foundation have picked up on this 'brain = innate' tendency, too. In response to an article in the *New York Times* that claimed from an fMRI study that 'a mother's impulse to love and protect her child appears to be hard-wired into her brain' one neuro-curmudgeon put out a plea to 'take experience and learning seriously. Just because you see a response [in the brain] – you don't get to claim it's hard-wired.'[9]

Another draw of neurononsense is what Yale researchers have referred to as 'the seductive allure of neuroscience explanations'. Deena Skolnick Weisberg and her colleagues found that people are pretty good at spotting bad explanations of psychological phenomena. Suppose, for example, you read about a study in which researchers found that men performed better than women on spatial reasoning tasks. Would you be satisfied by the circular explanation that 'women's poor performance relative to men's explains the gender difference in spatial reasoning abilities'? Probably not. The researchers aren't explaining their result, they're redescribing it: *women are worse at spatial reasoning because women are worse at spatial reasoning.* But simply add neuroscience and the same non-explanations suddenly seem much more satisfying:

> Brain scans of the right premotor area, known to be involved in spatial relational tasks, indicate that **women's poor performance relative to men's** causes different types of brain responses. This **explains the gender difference in spatial reasoning abilities**.

In bold text is the circular explanation that people found unsatisfying. The extra neuroscience bit tells us that spatial reasoning recruits a part of the brain, which should hardly surprise us. But it doesn't tell us *why* women performed worse than men. The explanation is still circular. But the neuroscience disguises this, even for students enrolled in an introductory cognitive science class, Weisberg and colleagues found.[10] Although it's not yet clear what it is, exactly, about neuroscience that is so persuasive, it's been found that people find scientific arguments more compelling when accompanied by an image showing brain activation rather than, say, a bar graph showing the same information.[11]

All of which should make us very concerned that this talk of brain differences might influence opinion and policy far more than it should. As Weisberg suggests, the seductive nature of neuroscience creates 'a dangerous situation in which it may not be the best research that wins debates in the public sphere.'[12]

The effects of neuroscience may be personal as well as political. Gender stereotypes are legitimated by these pseudo-scientific explanations. Suddenly, one is being modern and scientific, rather than old-fashioned and sexist. Do you want to claim, in a book for teachers and parents, that 'the world of the abstract . . . is explored more by the male brain than the female', thus explaining males' dominance in physics?[13] Why then, go right ahead! So long as the magic word *brain* is there, no further information required. But we have to wonder about the effect of this kind of information as it feeds back into society. As we saw in the first part of this book, the activation of gender stereotypes, even by means as subtle as our suspicion that they have found a home in the minds of others, can have measurable effects on our attitudes, identity and performance.

Neurosexism may also effect such changes directly. We can currently only speculate on the enervating effect of popular gender-science books on male patterns of leaving the milk to be bought by someone else. But there is evidence that media reports of gender that emphasise biological factors leave us more inclined to agree with gender stereotypes, to self-stereotype ourselves and even for our performance to fall in line with those stereotypes.[14] For example, one study found that women given a journal article to read that claimed that men are better at maths because of innate, biological and genetic differences performed worse on a GRE-like maths test than women shown an essay saying that men's greater effort underlies their superior performance. Likewise, women who had just read an essay arguing that there are genetically caused sex differences in mathematical ability performed substantially worse on a GRE-like test, compared with women who read that experiential factors explain sex differences in maths ability, psychologists Ilan Dar-Nimrod and Steven Heine found. (Being told this information by the experimenter had the same effect.) This damaging effect of the genetic account, the researchers suggest, may stem from people's assumption that genetically based differences are more profound and immutable than differences that arise from social factors. '[M]erely considering the role of genes in maths performance can have some deleterious consequences', they conclude. 'These findings raise discomforting questions regarding the effects that scientific theories can have on those who learn about them and the obligation that scientists have to be mindful of how their work is interpreted.'[15]

'Caveat lector' is Weisberg's advice. Neuroscientists who work in this area have some responsibility for how their findings of sex differences in the brain are interpreted and communicated. When this is done carelessly, it may have a real and significant impact on people's lives. Many neuroscientists do appear to be aware of this. They are appropriately cautious about interpreting sex differences in the brain, and many also take the time to remind journalists of just how far we are from mapping sex differences in the brain onto the mind. (And of course they may find their work being misrepresented, regardless.) Others, however, as we have seen, are more cavalier.

Not everyone would agree that the topic of sex differences in the brain requires a particular sensitivity. For example, sex-difference researcher Doreen Kimura has argued that '[w]e can't allow ourselves to get into a situation in which we say . . . "This is a finding that won't upset anyone, so I'm willing to generalize from it, but this other finding may be unpopular, so I need more evidence to support it before reporting it."'[16] I am not inclined to agree that the content of the research makes no difference to the degree of care scientists should take in generalising a result, or their concern in how it is popularised by others. I have, for example, heard neuroscientists who work in the area of drug dependency talk about the efforts they go to to prevent simplification or distortion of their findings by the media. This is not because they are worried about 'upsetting' people, but because it is a sensitive area, and 'brain facts' about dependency can change people's attitudes and feelings about a particular social group. These neuroscientists didn't seem to consider it unreasonable to work under a heavier burden of caution – a burden that I suggest it is also appropriate to place on those who comment on sex differences in the brain.[17]

Finally, there's an urgent need for editors, journalists and schools to develop far more sceptical attitudes towards claims made about sex differences in the brain. It is appalling to me that one can, apparently, say whatever drivel one likes about the male and female brain, and enjoy the pleasure of seeing it published in a reputable newspaper, changing a school's educational policy or becoming a best seller. Scientists can help here (as many already are). Weisberg suggests (in relation to the interpretation of imaging studies in general) that we 'take a more active stance as scientists, medical practitioners, and researchers.' She advocates that researchers become 'vocal critics' of misleading articles, put more pressure on 'newspaper and magazine writers to cover scientific issues with more depth and nuance', and, to this end, offer their expertise to members of the media.[18]

Neurosexism promotes damaging, limiting, potentially self-fulfilling stereotypes. Three years ago, I discovered my son's kindergarten teacher reading a book that claimed that his brain was incapable of forging the connection between emotion and language. And so I decided to write this book.[19] To make this kind of confident claim about hardwired psychological differences between males and females is to overlook the likelihood of spurious findings, the teething problems of new technology, the obscurity of the relationship between brain structure and psychological function and the difficulty of inferring psychological states from neuroimaging data. Dazzled by the seductive scientificness of neuroscience, commentators become blind to low-tech behavioural evidence of gender similarity, or flexibility in response to the social context. And, as we'll explore more in the next chapter, the very concept of hardwiring needs some updating.

A member of my family, who shall remain nameless, refers to all newborns as 'blobs'. There's a certain, limited truth to the description. Certainly, research continues to reveal just how sophisticated the neonate mind really is: already tuned to prefer its mother tongue, seek out facelike stimuli, time its waking up to coincide precisely with when its parents have just fallen most deeply into sleep. But it would not be an overstatement to say that newborns still have much to learn. Ideas about how this happens have been changing in important ways in neuroscience.

For decades, brain development has been thought of as an orderly adding in of new wiring that enables you to perform ever-more-sophisticated cognitive functions. According to this maturational viewpoint, gene activity at the appropriate time (and with the necessary experience and environment) brings about the maturation of new bits of neural circuitry. These are added in, enabling the child to reach new developmental milestones. Everyone, of course, acknowledges the essential role of experience on development. But when we think of brain development as a gene-directed process of adding in new circuitry, it's not difficult to see how the concept of hardwiring took off. It's been helped along by the popularity of evolutionary psychology, versions of which have promoted the idea that we are the luckless owners of seriously outdated neural circuitry that has been shaped by natural selection to match the environment of our hunter-gatherer ancestors.

But our brains, as we are now coming to understand, are

changed by our behaviour, our thinking, our social world. The new neuroconstructivist perspective of brain development emphasises the sheer exhilarating tangle of a continuous interaction among genes, brain and environment. Yes, gene expression gives rise to neural structures, and genetic material is itself impervious to outside influence. When it comes to genes, you get what you get. But gene *activity* is another story: genes switch on and off depending on what else is going on. Our environment, our behaviour, even our thinking, can all change what genes are *expressed*.[1] And thinking, learning, sensing can all change neural structure directly. As Bruce Wexler has argued, one important implication of this neuroplasticity is that we're not locked into the obsolete hardware of our ancestors:

> In addition to having the longest period during which brain growth is shaped by the environment, human beings alter the environment that shapes their brains to a degree without precedent among animals. . . . It is this ability to shape the environment that in turn shapes our brains that has allowed human adaptability and capability to develop at a much faster rate than is possible through alteration of the genetic code itself. This transgenerational shaping of brain function through culture also means that processes that govern the evolution of societies and cultures have a great influence on how our individual brains and minds work.[2]

It's important to point out that this is not a starry-eyed, environmentalist, we-can-all-be-anything-we-want-to-be viewpoint. Genes don't determine our brains (or our bodies), but they do constrain them. The developmental possibilities for an individual are neither infinitely malleable nor solely in the hands of the environment. But the insight that thinking, behaviour and experiences change the brain, directly, or through changes in genetic activity, seems to strip the word 'hardwiring' of much useful meaning. As neurophysiologist Ruth Bleier put it over two decades ago, we should 'view biology as potential, as capacity and not as static entity. Biology itself is socially influenced and defined; it changes and develops in interaction with and response to our minds and environment, as our behaviors do. Biology can be said to define possibilities but not determine them; it is never irrelevant but it is also not determinant.'[3]

And so, what do popular writers, scientists and former presidents of Harvard *mean* when they refer to gender differences as 'hardwired', or 'innate', or 'intrinsic', or 'inherent'? Some philosophers of biology, so far as I can tell, devote entire careers to the concept of innateness and what, if anything, it might mean. As cognitive neuroscientist Giordana Grossi points out, terms like *hardwired* – on loan from computer science where it refers to fixedness – translate poorly to the domain of neural circuits that change and learn throughout life, indeed, in *response* to life.[4]

Certainly, there is far more acknowledgement now of the role of experience and environment compared with a century or so ago. In the early twentieth century, '[g]enius was considered an innate quality which would naturally be manifested if it were possessed', as psychologist Stephanie Shields summarises.[5] No one now, I should think, would agree with this. And yet there remains, in some quarters, a Victorian-style attachment to notions of innate, immutable, inevitable qualities. How else to explain why the Greater Male Variability hypothesis – the idea that men are more likely to be outliers, good or bad ('more prodigies, more idiots'[6]) – appears to be no less appealing now than it was over a century ago?[7] In the early twentieth century, the Greater Male Variability hypothesis offered a neat explanation of why men so outnumbered women in eminence, despite the fact that there was little sex difference in the average scores of men and women on psychological tests. As Edward Thorndike (the sociologically unimaginative psychologist we met in the Introduction) explained it in 1910:

In particular, if men differ in intelligence and energy by wider extremes than do women, eminence in and leadership of the world's affairs of whatever sort will inevitably belong oftener to men. They will oftener deserve it.[8]

And today, it seems, they oftener deserve high-ranking positions in mathematics and science, according to Lawrence Summers:

> It does appear that on many, many different human attributes – height, weight, propensity for criminality, overall IQ, mathematical ability, scientific ability . . . there is a difference in the standard deviation and variability [statistical measures of the spread of a population] of a male and a female population. And that is true with respect to attributes that are and are not plausibly, culturally determined. If one supposes, as I think is reasonable, that if one is talking about physicists at a top twenty-five research university . . . small differences in the standard deviation will translate into very large differences in the available pool . . .[9]

I'd love to know, by the way, how extreme *non*criminality manifests itself. (Number of Supreme Court judges, perhaps?) But more to the point, the assertion that males are more variable in all regards – whether you're talking weight, height or SAT scores – certainly helps to frame variability as 'a guy thing' across the board. The implication is that there is something *inevitable* and immutable about greater male variability in mathematical and scientific ability. Certainly, in the furor that followed, Steven Pinker defended the idea of the timeless, universal nature of greater male variability ('biologists since Darwin have noted that for many traits and many species, males are the more variable gender').[10] Susan Pinker also plays the argument that '[m]en are simply more variable' in the shadow of the Summers controversy.[11] Her book displays a graph showing the findings from a report published by psychologist Ian Deary and his colleagues – a massive IQ study of 80,000 Scottish children born in 1921. Boys' and girls' average IQs were the same, the study found, but the boys' scores were more variable. But as the educational psychologist Leta Stetter Hollingworth pointed out in 1914, and as Ian Deary and his colleagues felt compelled to reiterate nearly 100 years later, 'the existence of sex differences either in means or variances in ability says nothing about the source or inevitability of such differences or their potential basis in immutable biology.'[12] This should be more obvious to us now than it was a hundred years ago when capacity for eminence was regarded as something that was simply 'in there'. We realise that, as Grossi has pointed out, '[m]athematics and science are learned in a period of time that spans across several years; passion and application need to be constantly nurtured and encouraged.'[13]

And, as it turns out, contemporary investigations of variability – both in the general population and in the most intellectually blessed pockets – have been showing that 'inevitable' and 'immutable' are adjectives that need not apply when it comes to describing greater male variability in mental ability. One cross-cultural study, published several years before the Summers debacle, compared sex differences in variability in verbal, maths and spatial abilities to see if the greater male variability in the United States was invariably seen in other countries. It was not. In each cognitive domain, there were countries in which females' scores were more variable than males'.[14]

More recently, several very large-scale studies have collected data that offer tests of the Greater Male Variability hypothesis by investigating whether males are inevitably more variable in maths performance, and always outnumber females at the high end of ability. The answer, in children at least, is no. In a *Science* study of over 7 million United States schoolchildren, Janet Hyde and her team found that across grade levels and states, boys were modestly more variable than girls. Yet when they looked at the data from Minnesota state assessments of eleventh graders to see how many boys and girls scored above the 95th and 99th percentile (that is, scored better than 95 percent, or 99 percent, of their peers) an interesting pattern emerged. Among white children there were, respectively, about one-and-a-half and two boys for every girl.

But among Asian American kids, the pattern was different. At the 95th percentile boys' advantage was less, and at the 99th percentile there were more girls than boys.[15] Start to look in other countries and you find further evidence that sex differences in variability are, well, variable. Luigi Guiso's cross-cultural *Science* study also found that, like the gender gap in mean scores, the ratio of males to females at the high end of performance is something that changes from country to country. While in the majority of the forty countries studied there were indeed more boys than girls at the 95th and 99th percentiles, in four countries the ratios were equal or even reversed. (These were Indonesia, the UK, Iceland and Thailand.)[16] Two other large cross-cultural studies of maths scores in teenagers have also found that although males are usually more variable, and outnumber girls at the top 5 percent of ability, this is not inevitably so: in some countries females are equally or more variable, or are as likely as boys to make it into the 95th percentile.[17]

Of course, scoring better than 95 or 99 percent of your school peers in mathematical ability is probably a baseline condition for eventually becoming a tenured Harvard professor of mathematics: like having hands, if you want to be a hairdresser. Top scorers on standardised maths tests may be what one group of researchers, rather stingily, refers to as 'the merely gifted'.[18] But also changeable is the proportion of girls identified in what's called the Study of Mathematically Precocious Youth (SMPY), which gives the quantitative section of the Scholastic Aptitude Test (the SAT) to kids who, theoretically, are way too young to take it. Children who score at least 700 (on a 200 to 800 scale) are defined as 'highly gifted'. In the early 1980s, highly gifted boys identified by the SMPY outnumbered girls 13 to 1. By 2005, this ratio had plummeted to 2.8 to 1.[19] That's a big change.

Being highly gifted is, I imagine, rather nice, but at the risk of swelling the head of any research mathematicians in top-ranked institutions who happen to be reading this book, they need to have made it onto the next rung of the giftedness ladder, and be 'profoundly gifted'. And here again – in this literally one-in-a-million category – there can be striking differences in female representation, depending on time, place and cultural background. The International Mathematical Olympiad (IMO) is a nine-hour exam, taken by six-person teams sent from up to ninety-five countries. The length of the exam is off-putting enough, but the six problems within it are also so difficult that every year just a few students (or sometimes even none) get a perfect score. We tend not to hear that much about maths competitions (perhaps in part because, let's be honest, live televised coverage of a nine-hour maths exam would not make for compelling viewing). So it's probably worth pointing out that these competitions are not female-free zones. Girls are among those who achieve perfect scores. Girls, like US team member Sherry Gong, win medals for outstanding performance. Gong won a silver medal in the 2005 IMO and a gold medal in 2007. The girl can do maths – and she's not alone. As the researchers point out, 'numerous girls exist who possess truly profound ability in mathematical problem solving.'[20]

But an equally important insight from their analysis is what a difference where you come from makes for your chances of being identified and nurtured as a maths whiz. Between 1998 and 2008 *no* girls competed for Japan. But next door, seven girls competed for South Korea (which, by the way, ranks higher than Japan). A profoundly gifted young female mathematician in Slovakia has a five times greater chance of being included on the IMO team than her counterpart in the neighbouring Czech Republic. (Again, Slovakia outperforms the Czech Republic. I say this not to be competitive, but merely to show that teams with more girls have not been scraping the bottom of the barrel.) The ratio of female members on IMO teams among the top 34 participating countries ranges from none at all, to 1 in 4 (in Serbia and Montenegro). This is not random fluctuation, but evidence of 'socio-cultural, educational, or other environmental factors' at work.[21]

In fact, we can see this very clearly even within North America. Being underrepresented on the IMO team, or the Mathematical

Olympiad Summer Program (MOSP), is not, as you might assume, a *girl* problem. It's more subtle and interesting than that. First of all, if you're Hispanic, African American or Native American, it matters not whether you have two X chromosomes or one – you might as well give up now on any dreams of sweating for nine hours over some proofs. Then within girls, interesting patterns emerge. Asian American girls are *not* underrepresented, relative to their numbers in the population. But that doesn't mean that it's even simply a *white girl* problem. Non-Hispanic white girls born in North America are sorely underrepresented: there are about twenty times fewer of them on IMO teams than you'd expect based on their numbers in the population, and they virtually never attend the highly selective MOSP. But this isn't the case for non-Hispanic white girls who were born in Europe, immigrants from countries like Romania, Russia and the Ukraine, who manage on the whole to keep their end up when it comes to participating in these prestigious competitions and programmes. The success of this group of women continues into their careers. These women are *a hundred times more likely* to make it into the maths faculty of Harvard, MIT, Princeton, Stanford or University of California–Berkeley than their native-born white counterparts. They do every bit as well as white males, relative to their numbers in the population. As the researchers conclude:

> Taken together, these data indicate that the scarcity of USA and Canadian girl IMO participants is probably due, in significant part, to socio-cultural and other environmental factors, not race or gender *per se*. These factors likely inhibit native-born white and historically underrepresented minority girls with exceptional mathematical talent from being identified and nurtured to excel in mathematics. Assuming environmental factors inhibit most mathematically gifted girls being raised in most cultures in most countries at most times from pursuing mathematics to the best of their ability, we estimate the *lower* bound on the percentage of children

with IMO medal-level mathematical talent who are girls to be in the 12%–24% range [i.e., the levels seen in countries like Serbia and Montenegro]. . . . In a gender-neutral society, the real percentage could be significantly higher; however, we currently lack ways to measure it.[22]

That's a lot of squandered talent, and among boys, too. As the researchers acknowledge, the data they collected can't answer the question of whether females – in a perfectly gender-equal environment – could match (or, why not be bold, perhaps even surpass) males in maths. But the gender gap is narrowing all the time, and shows that mathematical eminence is not fixed, or hardwired or intrinsic, but is instead responsive to cultural factors that affect the extent to which mathematical talent is identified and nurtured, or passed over, stifled or suppressed in males and females.

And so this is all good news for Lawrence Summers, who said that he 'would far prefer to believe something else' than the 'unfortunate truth' that, in part, 'differing variances' lie behind women's underrepresentation in science.[23] And for Pinker, too, who warned Summers' detractors that '[h]istory tells us that how much we want to believe a proposition is not a reliable guide as to whether it is true.'[24] Evidence for the malleability of the gender gap in ability and achievement is there. And this is important because, as we learned in the first part of the book, it makes a difference what we believe about difference. Stanford University's psychologist Carol Dweck and her colleagues have discovered that what you believe about intellectual ability – whether you think it's a fixed gift, or an earned quality that can be developed – makes a difference to your behaviour, persistence and performance. Students who see ability as fixed – a gift – are more vulnerable to setbacks and difficulties. And stereotypes, as Dweck rightly points out, 'are stories about gifts – about who has them and who doesn't.'[25] Dweck and her colleagues have shown that when students are encouraged to see maths ability as something that grows with effort – pointing out, for example, that the brain forges new connections and develops better ability every time they practise a task – grades improve and gender gaps diminish (relative to groups given control interventions).[26] The Greater Male Variability hypothesis, of course, endorses the view that very great intellectual ability is indeed a fixed trait, a gift bestowed almost exclusively on men. Add a little talk of women's insufficient white matter volumes, or their plump corpora callosa, and the ingredients for a self-fulfilling prophecy are all in place.

The sensitivity of the mind to neuroscientific claims about difference raises ethical concerns.[27] A recent study by University of Exeter psychologist Thomas Morton and his colleagues asked one group of participants to read the kind of passage that is the bread-and-butter of a certain type of popular gender science book. It presented essentialist theories – that gender difference in thinking and behaviour are biological, stable and immutable – as scientifically established facts. A second group read a similar article, but one in which the claims were presented as being under debate in the scientific community. The 'fact' article led people to more strongly endorse biological theories of gender difference, to be more confident that society treats women fairly and to feel less certain that the gender status quo is likely to change. It also left men rather more cavalier about discriminatory practices: compared with men who read the 'debate' article, they agreed more with statements like, 'If I would work in a company where my manager preferred hiring men to women, I would privately support him', and 'If I were a manager in a company myself, I would believe that more often than not, promoting men is a better investment in the future of the company than promoting women.' They also felt better about themselves – a small consolation indeed to women, I think you'll agree.

Interestingly, for men who tend to the view that sex

discrimination is a thing of the past, the appeal of essentialist research is enhanced by evidence that the gender gap is closing, Morton and his colleagues also found. Participants were asked to rate research that investigated the genetic basis of sex differences in mouse brains, as well as claiming that similar factors may underlie psychological gender differences in humans. Beforehand they read an article, supposedly from a national newspaper, arguing either that gender inequality was stable, or closing. After reading about women's gains these men more readily agreed that 'this type of research should continue, deserved more funding, was good for society, represented the facts about gender differences, and made a major contribution to understanding human nature'.[28]

Taken together, Morton's findings suggest that women's gains will, in certain quarters, increase demand for essentialist research. As this research trickles back into society, people will turn away from social and structural explanations of gender difference. They will give up on the idea of further social change. And, to help the belief in the inevitability of inequality come true, workplace discrimination against women will increase.

It is, I think, time to raise the bar when it comes to the interpretation and communication of sex differences in the brain. How long, exactly, do we need to learn from the mistakes of the past?

As we've seen in this part of the book, speculating about sex differences from the frontiers of science is not a job for the fainthearted who hate to get it wrong. So far, the items on that list of brain differences that are thought to explain the gender status quo have always, in the end, been crossed off.[29] But before this happens, speculation becomes elevated to the status of fact, especially in the hands of some popular writers. Once in the public domain these supposed facts about male and female brains become part of the culture, often lingering on well past their best-by dates. Here, they reinforce and legitimate the gender stereotypes that interact with our minds, helping to create the very gender inequalities that the neuroscientific claims seek to explain.[30]

PART 3

Recycling Gender

'It's made me think a lot more about genetic influence, she's got two X chromosomes, and that somehow, I don't know, because we don't push the Barbie stuff at all, in fact I would prefer her not to have it . . . so I'm kind of intrigued at how even though I am sort of doing the middle of the road, that she is nonetheless veering over towards being more feminine, and I think it's genetic.' (White, upper-middle class, lesbian mother, describing her three-year-old daughter).

—Comment from Emily Kane's interview study (2006)

When I tell parents that I'm writing a book about gender, the most common response I get is an anecdote about how they tried gender-neutral parenting, and it simply didn't work. (The next most frequent reaction is a polite edging away.) This is a common experience, found sociologist Emily Kane. She interviewed forty-two parents of preschoolers, from a wide range of backgrounds, and asked them why they thought that their sons or daughters sometimes behaved in sex-typed ways. Many parents called on evolutionary or divine reasons to explain why there should be innate biological differences between girls and boys (although most also mentioned social factors). But over a third of the interviewed parents – mostly white and middle or upper middle class – expressed the 'biology as fallback' position, as Kane called it. Only by process of elimination did they come to the conclusion that differences between boys and girls were biological. Believing that they practised gender-neutral parenting, biology was the only remaining explanation:

> 'It's not as if (my sons) haven't been exposed to all that princess stuff . . . they're around it, but they show no interest, they haven't been clamouring for any special princess toys or Ken and Barbie stuff . . . I think that's the hard-wired stuff, to even see it and for it to be prevalent, and to not be interested in it.' (White, upper-middle class, heterosexual father, describing his three and four year old sons' lack of interest in their six-year-old sister's toys).

Parents see their young children behaving in stereotypically boyish or girlish ways and, as Kane puts it, 'assume that only something immutable could intervene between their gender-neutral efforts and the gendered outcomes they witness.'[1]

They are in distinguished company. As part of his suggestions regarding women's possible intrinsically inferior aptitude for, and interest in, high-level scientific careers, Lawrence Summers offered an opinion on the essential differences between the sexes, gleaned from the nursery hearth:

> So, I think, while I would prefer to believe otherwise, I guess my experience with my two and a half year old twin daughters who were not given dolls and who were given trucks, and found themselves saying to each other, look, daddy truck is carrying the baby truck, tells me something. And I think it's just something that you probably have to recognize.[2]

Likewise, in a scientific debate about the reasons behind the gender gap in science, Steven Pinker joked: 'It is said that there is a technical term for people who believe that little boys and little girls are born indistinguishable and are moulded into their natures by parental socialization. The term is "childless".'[3]

The frustration of the naively nonsexist parent has become a staple joke. An all but obligatory paragraph in contemporary books and articles about hardwired gender differences gleefully describes a parent's valiant, but always comically hopeless, attempts at gender-neutral parenting:

> One of my [Louann Brizendine's] patients gave her three-and-a-half-year-old daughter many unisex toys, including a bright red fire truck instead of a doll. She walked into her daughter's room one afternoon to find her cuddling the truck in a baby blanket, rocking it back and forth saying, 'Don't worry, little truckie, everything will be all right.'[4]

As it happens, I can match anecdote with counter anecdote. Both of my sons, as toddlers, behaved in much the same way as Lawrence Summers's and Brizendine's patient's young daughters. They too, despite being male, tucked trucks into pretend beds and, yes, called them Daddy, Mummy and Baby.

Yet parents are right when they say that young boys and girls play differently, even if the contrast isn't nearly as black-and-white as it's often portrayed. As the quotations with which this chapter began suggest, the received popular wisdom is that this happens despite the nonsexist, gender-neutral environment in which children are now raised: 'Today we know that the truth is . . . [that] parents raise girls and boys differently because girls and boys are so different from birth. Girls and boys behave differently because their brains are wired differently', says Leonard Sax.[5]

Well, as we now know, there's more than one loophole in the 'wiring' argument. And as we'll see in this part of the book, there are many reasons, ranging from subtle to blatant, why a gender-neutral environment is not something that any parent does, could or perhaps even wants to provide.

The obstacles to gender-neutral parenting begin well before a baby is born. When Emily Kane asked her sample of parents about their preferences for sons or daughters before they even became parents, the themes of their responses showed that they had gendered expectations of even hypothetical children. The men tended to want a son, a common reason being that they liked the idea of teaching him to play sports. 'I always wanted a son . . . I think that's just a normal thing for a guy to want. I wanted to teach my son to play basketball, I wanted to teach my son to play baseball, and so forth. Just thinking of all the things you could do with your son' was how one father put it. (An alien researcher from outer space, reading Kane's transcripts, might be forgiven for coming to the conclusion that human females are born without arms and legs.) Mothers in the study, too, seemed to fall in with the assumption that boys and girls are good for different things. Kane found that if mothers wanted a son, it was to provide their husbands with a companion with whom to do things, like sports, that apparently couldn't be done with girls. Daughters, by contrast, were expected to offer very different kinds of parental experiences: 'A girl, I wanted that more . . . to dress her up and to buy the dolls and you know, the dance classes . . . A girl was someone that you could do all the things that you like to do with more than you could a boy.' More often, though, girls were wanted because of the emotional connection they would provide. Only a daughter would be naturally inclined to emotional intimacy and the remembering of birthdays, was the unspoken assumption. Not yet conceived, and already the sons were off the hook for remembering to call or send birthday flowers.[6]

Postconception, the gendered expectations continue. Sociologist Barbara Rothman asked a group of mothers to describe the movements of their foetuses in the last three months of pregnancy. Among the women who didn't know the sex of their baby while they were pregnant, there was no particular pattern to the way that (what turned out to be) male and female babies were described. But women who knew the sex of their unborn baby described the movements of sons and daughters differently. All were 'active', but male activity was more likely to be described as 'vigorous' and 'strong', including what Rothman teasingly describes as 'the "John Wayne fetus" – "calm but strong"'. Female activity, by contrast, was described in gentler terms: '*Not* violent, *not* excessively energetic, *not* terribly active were used for females'.[7]

Then, there are the intriguing experiences of Kara Smith, an educational researcher with a background in women's studies, who kept pregnancy field notes. Throughout the entire nine months of the pregnancy, Smith noted all the words and feelings expressed to the unborn baby. And, in the sixth month of the pregnancy, an ultrasound revealed his sex:

> He was a boy. He was 'stronger' now than the child I had known only one minute before. He did not need to be addressed with such light and fluffy language, such as 'little one'. . . . Thus, I lowered my voice to a deeper octave. It lost its tenderness. The tone in my voice was more articulate and short, whereas, before, the pitch in my voice was high and feminine. I wanted him to be 'strong' and 'athletic', therefore, I had to speak to him with a stereotypical 'strong', 'masculine' voice to encourage this 'innate strength'.

What startled Smith most about this exercise was that someone like herself, well-versed in the negative consequences of gender socialisation, was inadvertently drawing on stereotypes in the way she responded to the baby. 'I was, quite honestly, shocked by the findings', she writes. Here was a mother – and, let's not forget, not just any old mother, but the sort of feminist mother so beloved of unisex-parenting-gone-wrong stories – finding herself socialising her child into gender roles before he was even born.[8]

This is just one person's experience. But Smith's observation – that her behaviour was undermining her values – is backed by a large body of research. If all of our actions and judgements stemmed from reflected, consciously endorsed beliefs and values then not only would the world be a better place, but this book would be several pages shorter. Social psychologists, who have been unravelling how implicit and explicit processes interact to make up our perceptions, feelings and behaviour, stress the importance of understanding 'what happens in minds without explicit permission.'[9] And this is particularly important when implicit associations don't match the more-modern beliefs of the conscious mind. Implicit attitudes play an important part in our psychology. They distort social perception, they leak out into our behaviour, they influence our decisions – and all without us realising.[10]

Parents' gender associations are firmly in place well before a child is even a twinkle in daddy's eye. The scant but suggestive data of this chapter hint that beliefs about gender – either consciously or unconsciously held – are already shaping expectations about a future child's interests and values, already biasing the mother's perception of the little kicking baby inside her, and are already moulding a mother's communication with her unborn child.

And then, the baby is born.

It's a Boy! *'Rob and Kris are thrilled to announce the safe arrival of Jack Morgan Tinker. Proud grandparents are Hollis and Marilyn Clifton of Ottawa and Larry and Rosemary Tinker of Montreal. Welcome little one!'* It's a Girl! *'Barbara Lofton and Scott Hasler are delighted to announce the birth of their lovely daughter, Madison Evelyn Hasler. Grandparents are both joyful and overwhelmed.'*

You can learn a lot from birth announcements. In 2004, McGill University researchers analysed nearly 400 birth announcements placed by parents in two Canadian newspapers, and examined them for expressions of happiness and pride. Parents of boys, they found, expressed more pride in the news, while parents of girls expressed greater happiness. Why would parents officially report different emotional reactions to the birth of a boy versus a girl? The authors suggest that the birth of a girl more powerfully triggers the warm, fuzzy feelings relating to attachment, while the greater pride in a boy stems from an unconscious belief that a boy will enhance standing in the social world.[11]

Parents may also be slightly more likely to place a birth announcement for a boy than for a girl, discovered psychologist John Jost and his colleagues. Male births make up 51 percent of live births, so one would expect the same percentage of birth announcements to be for boys. However, in their data set of thousands of Florida birth announcements, more were for male babies than one would expect: 53 percent. It's a very small (although statistically significant) difference, it's true. (And it only held for traditional families, in which the mother had taken on the father's last name.) But as the authors point out, '[t]he fact that gender differences show up at all for a family decision that is such a clear and significant reflection of parental pride is both surprising and worrisome. We suspect that most parents would be shocked and embarrassed to learn that they might have publicly announced the birth of a son, but not a daughter, and this suggests that the effect is subtle, implicit, and yet powerful.'[12] Not so long ago in Western societies, males were quite openly valued over females (and this is still the case in many developing countries). Today, we don't think one sex better or more valuable than the other – and yet, at an implicit level, could we still be holding males in higher regard?

A close look at the names given to the babies in this data set suggested that we might. Jost and his colleagues also analysed the thousands of birth announcements to see how often sons and daughters were given a name that began with the same letter as either the father's or mother's name: for example, Russell and Karen calling their son Rory versus Kevin. How, you may well wonder, does this exercise reveal anything at all about the machinations of the implicit mind? The reason is that, remarkably, not all letters of the alphabet are equal in the eye of the beholder. People unconsciously place a special value on the letter that begins their own name. With this phenomenon in mind, Jost and colleagues looked for evidence of 'implicit paternalism' in the names that parents chose for their children. They found that boys were more likely to be given names that began with the paternal first initial than the maternal initial, but girls were equally likely to share a first initial with their mother or father. (And this wasn't because of sons being named after their dads; kids with exactly the same name were excluded from this analysis.) In other words, parents seemed to be unconsciously overvaluing fathers' names and perhaps also boys, who were more often bestowed the higher-value male initial.[13]

Clearly, naming a child is a highly personal, multifaceted process. It's impossible to know for sure what is behind these surprising findings. But as Jost and colleagues point out, contemporary manifestations of sexism and racism are often 'indirect, subtle, and (in some cases) non-conscious.'[14] In modern, developed societies, males and females are legally – and no doubt also in the eyes of most parents – born with equal status and entitled to the same opportunities. Yet of course this egalitarian attitude is very new, and it's poorly reflected in the distribution of political, social, economic and sometimes even personal power between the sexes. It's a 'half-changed world', as Peggy Orenstein put it[15] and here, in the naming of children and composing of birth announcements, are little strands of evidence of parents' half-changed minds. Without meaning to, and without realising it, we may be valuing boys and girls differently, and for different qualities, within hours of birth.

From this starting point, unequal even before conception, parenting begins.

When psychologists run experiments in search of differences between boy babies and girl babies, they do not order in unused babies still in their shrink-wrapped packages. Even newborns show a preference for their native language, presumably from hearing, in utero, the intonation and rhythm of their mother tongue.[1] Babies are button-nosed little learning machines. For example, developmental psychologist Paul Quinn and his colleagues found that babies just three to four months of age prefer to look at female, rather than male, faces.[2] The researchers wondered whether this might be because the babies had spent most of their time with female caregivers and that greater familiarity with female faces was the reason they liked them more. And so they tested a small group of daddy-reared babies and found that *this* rare breed of baby preferred male faces. (A further experiment suggested that babies' preference for faces of the more familiar sex stems from acquired expertise with those kinds of faces.) Likewise, although they have no preference at birth, by three months of age, babies look more at same-race faces than other-race faces.[3] Babies are also, even in the first year of life, sensitive to the emotional reactions of caregivers. They use facial expressions and tone of voice as a guide to what toys, for example, should be approached and, especially, what should be avoided.[4] Interestingly, infants find mixed messages – even those that include some sort of positive expression towards a toy – somewhat off-putting.[5]

These sorts of discoveries mean we have to take babies' environments and experiences seriously when we try to understand any differences between even very young boys and girls. Of course, if parents provide a truly gender-neutral environment for their babies, then this won't matter. But do they?

Certainly, the physical environments of baby girls and boys are not identical. Without doubt, your typical baby girl has a lot more pink in her life, and a baby boy a great deal more blue. And they may also have different levels of exposure to dolls and trucks at even a very tender age. Alison Nash and Rosemary Krawczyk inventoried the toys of more than 200 children in New York and Minnesota. They found that even among six- to twelve-month-old infants, the youngest age group they studied, boys had more 'toys of the world' (like transportation vehicles and machines) while girls had more 'toys of the home' (like dolls and housekeeping toys).[6]

We can also justifiably wonder whether baby boys' and girls' psychological environments are the same. Psychologists often find that parents treat baby girls and boys differently, despite an absence of any discernible differences in the babies' behaviour or abilities. One study, for example, found that mothers conversed and interacted more with girl babies and young toddlers, even when they were as young as six months old.[7] This was despite the fact that boys were no less responsive to their mother's speech and were no more likely to leave their mother's side. As the authors suggest, this may help girls learn the higher level of social interaction expected of them, and boys the greater independence. Mothers are also more sensitive to changes in facial expressions of happiness when an unfamiliar six-month-old baby is labelled as a girl rather than a boy, suggesting that their gendered expectations affect their perception of babies' emotions.[8] Gendered expectations also seem to bias mothers' perception of their infants' physical abilities. Mothers were shown an adjustable sloping walkway, and asked to estimate the steepness of slope their crawling eleven-month-old child could manage and would attempt. Girls and boys differed in neither crawling ability nor risk taking when it came to testing

them on the walkway. But mothers underestimated girls and over-
estimated boys – both in crawling ability and crawling attempts
– meaning that in the real world they might often wrongly think
their daughters incapable of performing or attempting some motor
feats, and equally erroneously think their sons capable of others.[9]
As infants reach the toddler and preschool years, researchers find
that mothers talk more to girls than to boys, and that they talk
about emotions differently to the two sexes – and in a way that's
consistent with (and sometimes helps to create the truth of) the
stereotyped belief that females are the emotion experts.[10]

It seems, then, that gender stereotypes, even if perhaps only
implicitly held, affect parents' behaviour towards their babies. This
is hardly surprising. Implicit associations don't, after all, remain
carefully locked away in the unconscious. They can play an important part in behaviour and may tend to leak out when we aren't
thinking too much, or *can't* think too much, about what we are
doing – perhaps in our tone of voice, or body language. Implicit
attitudes can also take the upper hand when it comes to our behaviour when we are distracted, tired or under pressure of time (conditions that, from personal experience, I would estimate are fulfilled
about 99 percent of the time while parenting).[11] Is it possible that
parents' implicit attitudes about gender might be subtly transmit-
ted to their children?

Here is a transcript from a video clip shown to three- to six-
year-old children, by psychologist Luigi Castelli and his colleagues:

ABDUL [black adult male]: Hi, my name is Abdul and I come
    from Senegal which is an African country.
GASPARE [white adult male]: Hi, my name is Gaspare. I come
    from Padova. I'm Italian. I have nothing against the fact that
    people from other countries and, possibly, with a different
    colour of the skin, come and live in Italy with us. I'm happy
    if you come to live in our city. I believe we must be toler-
    ant and welcome everyone in the same way, and I do not
    really care about the colour of the skin. For instance, if my

child would become friends with a child whose skin is black
I would be very happy. In order to live in a better world we
must overcome the differences between us.

When it comes to holding a generous, open-armed policy towards
people with different skin tones Gaspare, I think we can all agree,
cannot be faulted. Psychologist Luigi Castelli and colleagues
showed two groups of preschoolers a video clip in which Gaspare
expressed these egalitarian, colour-blind opinions, and then asked
the children questions like *Would you like to play with Abdul?* or
How much do you like Abdul? A third and fourth group of children
were asked the same questions after seeing a slightly different clip.
In this alternative clip, Gaspare steered clear of race politics altogether, and talked only about his work in a dress shop.

So, which group of children felt most warmly towards Abdul?
Was it, as you might expect, the children who heard Gaspare's
positive, moving words about our common humanity? In fact, no.
It made no difference. But something else, unspoken, did.

In half of his positive speech clips, Gaspare's nonverbal behaviour matched his words: he shook Abdul's hand with enthusiasm;
he spoke enthusiastically; he sat near Abdul, leaned towards him,
and regularly looked right at him. But in the other positive speech
clip, Gaspare's actions belied his verbal sentiments: his handshake
was flaccid; his voice was slow and hesitant. Gaspare also kept
an empty seat between himself and Abdul, leaned away from his
African acquaintance, and avoided eye contact. Likewise, in the
verbally neutral clip, sometimes Gaspare's body language was
positive, and sometimes it was negative. It was these nonverbal
cues the children picked up on. To them, the nonverbal actions
spoke louder than words. Children who saw the enthusiastic physical behaviours – regardless of what Gaspare actually *said* – felt
significantly more friendly towards Abdul than children who saw
Gaspare's body express unease.[12]

To the researchers, this was no surprise, just another piece
of the puzzle of children's racial attitudes. It's natural to assume

that children, at least to some extent, pick up their views about other ethnic groups from their parents. And yet when you canvass parents and their children on this subject, their answers simply don't match up. More (or less) prejudiced parents don't have more (or less) prejudiced children, particularly at younger ages.[13] But that's when you just ask outright. Recently, however, Castelli and his colleagues found that white mothers' *implicit* race attitudes *do* match the racial attitudes of their offspring. Their consciously expressed attitudes seem to have no influence on the children. But the stronger the mother's implicit negativity towards black people (measured using the Implicit Association Test), the less likely her child is to choose a black child to play with, and to rate a black peer in a positive, charitable fashion.[14]

When it comes to race, children seem to be learning from the wrong half of the half-changed mind. That's not to say that children are oblivious to what is said. (For ethical reasons, the researchers didn't show a racist clip. As they point out, if they had used this as a contrast to the positive message they might well have seen a greater impact of the verbal message.) The point is that they also learn from what is *not* said, but expressed in other, more subtle ways, and even when this contradicts the spoken message. To my knowledge, no one has yet explored whether children's gender attitudes are influenced by a parent's implicit gender associations. But, intriguingly, there seems to be no relationship at all between parents' and children's explicit gender attitudes in those early preschool years.[15] Castelli's findings prick the suspicion that it is not that young children are learning nothing about gender from their parents, but are instead picking up on the gendered patterns of their parents' implicit minds. Is it possible, for example, that parents subtly and inadvertently convey ambivalence about cross-gender play – an unenthusiastic tone of voice, a withdrawing of attention – from which infants perceive and learn? As psychologists Nancy Weitzman and her colleagues suggested over twenty years ago, 'expressed attitudes may be easier to change than deeply entrenched, nonconscious forms of behavior'.[16] The research tools are now available for developmental psychologists to investigate how parents' implicit attitudes about gender affect their behaviour and their children, and it will be interesting to see what they find.

There are certainly more than a few signs that contemporary parents have mixed feelings about the very idea of successfully rearing unisex children. A large meta-analysis in 1991 gathered together all the studies that looked at whether parents treat boys and girls differently.[17] While in many ways parents seemed to treat boys and girls much the same, in one domain they clearly did not: parents encouraged gender-typed activities and play, and discouraged cross-gender behaviour. Of course, this study is now around two decades old and there are some indications that, these days, parents are actively encouraging cross-gender play. But, scratch the surface of these genuinely egalitarian values, and the contradictions of the half-changed mind still appear, especially for boys. The parents in a small study of twenty-six preschoolers from a southeastern city almost all agreed that girls should be encouraged to play with building blocks and toy trucks, and to play Little League and other competitive sports. However, when the researchers asked the children themselves whether their parents would approve of cross-gender play (*What would Mum think of that? Would Dad like you to play with one of those?*), they heard a rather different story. For instance, only a quarter of the three-year-old girls thought that their mother would want them to play with a baseball and mitt, or a skateboard (both of which the little girls readily identified as 'for boys'), compared with 80 percent of the three-year-old boys.

The same parents also all but unanimously thought it important for both boys and girls to develop social skills. Yet in apparent contradiction to this belief a third of them, when asked, were either uncertain whether they would buy their son a doll or would definitely not do so. Interestingly, the three- and five-year-old boys tested were well-aware of this ambivalence, with just two of the twelve boys of the opinion that their parents would be happy for them to play with a doll. That's a far cry from a gender-neutral environment.[18]

The parents interviewed by Emily Kane, by contrast, were more liberal (although we don't know how the children perceived their parents' attitudes). She found that these parents 'celebrated' and even encouraged gender nonconformity in their young daughters. 'I don't want her just to color and play with dolls, I want her to be athletic', one father said. They also mostly 'accepted, and often even celebrated' activities they thought would promote domestic skills, nurturance and empathy in their sons – including play with dolls, toy kitchens and tea sets (although sometimes this acceptance was rather grudging). However, even in these parents there was evidence that the gender border was being carefully negotiated and patrolled for boys. Many parents drew the line at Barbie, for instance (who was regularly requested by the little boys) or tried to diminish her quintessential femininity: 'I would ask him, "What do you want for your birthday?" . . . and he always kept saying Barbie . . . So we compromised, we got him a NASCAR [National Association for Stock Car Auto Racing] Barbie.' Another father said that if his son 'really wanted to dance, I'd let him . . ., but at the same time, I'd be doing other things to compensate for the fact that I signed him up for dance.'[19]

In curious contradiction to their explanations of their preschoolers' gender-stereotypical behaviours (many, you will recall, turning to biology as the only possible remaining explanation), Kane found it 'striking . . . how frequently parents indicated that they took action to craft an appropriate gender performance with and for their preschool-aged sons, viewing masculinity as something they needed to work on to accomplish.'[20] Cross-gender behaviour is seen as less acceptable in boys than it is in girls: unlike the term 'tomboy' there is nothing positive implied by its male counterpart, the 'sissy'.[21] Parents were aware of the backlash they might, or indeed *had*, received from others when they allowed their children to deviate from gender norms. '[P]arents [are] thinking consciously, even strategically, about their children's gender performance, and sometimes crafting it to ensure not their children's free agency but instead their structured and successful performance of gender', argues Kane.[22]

From these admittedly limited data, an interesting picture emerges. As Orenstein described the state of flux of the twenty-first century, '[o]ld patterns and expectations have broken down, but new ideas seem fragmentary, unrealistic, and often contradictory.'[23] Some parents, at least, genuinely want to rear children outside the constraints of rigid stereotypes, yet even before children are born parents have different expectations of them. They sincerely believe that boys and girls deserve to be free to develop their own interests and to become rounded individuals – gender norms be damned – yet at the same time they channel and craft their children's 'gender performances', especially for boys. (For girls, this pressure may kick in more during adolescence, some researchers suggest.) Parents say they are open-minded about their sons taking up nontraditional careers, like nursing – but in the very same questionnaire they reveal a preference that their sons behave in gender-typical ways. And, even though they sincerely claim to hold the two sexes as equal, parents simultaneously devalue the feminine and limit boys' access to it.

A parent with a half-changed mind (or perhaps even mostly *un*changed with an egalitarian veneer) will not parent in a spotlessly gender-neutral fashion. A parent who has just read an impressively scientific-sounding popular book or article about how boy and girl babies come differently prewired, or have differently structured brains, might not even try. Babies, in turn, seem to be primed to like what is familiar and are remarkably sensitive to their social world. So what, then, are we to make of recent evidence that children show gender-stereotyped interests before they are even two years old? For example, psychologist Gerianne Alexander and her colleagues measured how long five- to six-month-old babies looked at a pink dolly and a blue truck. There weren't any differences between boys and girls in how long they looked at each type of toy. But when the researchers counted up the number of times the babies briefly fixated on each toy (that is, when gaze remained still for at least 100 milliseconds), they found that girls were less interested in the truck: they fixated on it less than on the doll and less than did boys.[24] And at just one year of age – when offered cars, dolls, beauty sets and so on – boys and girls have been found to play in sex-stereotypical ways in the lab. One study, for instance, found that one-year-old boys played longer with boyish toys than did girls, while girls spent longer with girlish toys than did boys. At this age cross-gender toys haven't yet acquired a 'hot potato' quality and the differences in play behaviour are very modest.[25] Despite the gender differences seen in this particular study, for instance, boys still spent 37 percent of their total playing time with girlish toys (compared with 46 percent of their time with boyish toys).[26] Similarly, another study of one-year-olds found that, although boys this age played more with the boyish toys, the sexes spent a similar amount of time with girlish toys and were equally likely to choose a ball, a doll or a car as a gift from the experimenter.[27]

Still, there *are* differences, and at first glance these findings seem to toll the bell for the idea that children's gendered play preferences are purely socially constructed. The reason is that infants at this young age, so far as we know, are not aware of their own sex. They can't therefore be basing their behaviour on reasoning along the lines of *I am a girl and girls do not play with trucks*. Sax argues that the findings from this kind of research spell an end to the '"Dark Ages" – that period from the mid-1960s to the mid-1990s during which it was politically incorrect to suggest that there were innate differences in how girls and boys learn and play'.[28] Yet do these subtle differences reflect hardwired predispositions that differ between the sexes (a possibility that, by the way, developmental psychologists who are interested in social influences on play behaviour readily acknowledge)? Or do they reflect babies' sensitivity to their social and physical worlds? Does a six-month-old girl look longer at a pink doll than a blue truck because that's how she's wired or because she's seen more pink and more dolls in her short life (especially paired with pleasurable experiences with caregivers) and less blue and fewer trucks?[29] Does a one-year-old boy really play less with a plastic tea set because of hardwiring?[30] What are we to make of boys' greater interest in looking at balls and vehicles over feminine toys at nine months of age, given that six months earlier they looked at dolls, ovens and strollers just as much?[31] These are questions that deserve some thought.

Whether subtle (or even not-so-subtle) differences in the experiences, environments, toys, encouragement, and nonverbal communication offered to baby boys and girls can explain their modestly gendered early interests remains to be seen. Infants and toddlers don't need to know whether they are a boy or a girl to nonetheless be responsive to their parents' 'structuring, channeling, modeling, labeling, and reacting evaluatively to gender-linked conduct', as psychologists Albert Bandura and Kay Bussey have pointed out.[32]

But what is indisputable is that, as we'll see in the next chapter, we make the mystery of gender as easy as possible for children to solve.

If you're ever feeling bored and aimless in a shopping centre, try this experiment. Visit ten children's clothing stores, and each time approach a salesperson saying that you are looking for a present for a newborn. Count how many times you are asked, 'Is it a boy or a girl?' You are likely to have a 100 percent hit rate if you try this one spare afternoon. It is so ubiquitous now to dress and accessorise boys and girls differently, from birth, that it is easy to forget to wonder *why* we do this or to ask what children themselves might make of this rigidly adhered-to code. And it *is* a rigid code. I recently stood in a clothing store, paralysed with indecision as I deliberated which outfit to choose for a friend's new baby girl. The cutest one had little honking cars on it. Yet even though my friend lives in England, rather than Saudi Arabia, I just couldn't choose it. I knew that if my friend ever *did* put her baby in that outfit (rather than just toss it in the charity pile thinking, The sooner Cordelia finishes that book on gender the better . . .), she would spend the rest of the day correcting strangers who congratulated her on her beautiful baby boy. And well before dinnertime she would have learned that you can dress babies in clothing intended for the other sex or you can avoid being looked at as if you were insane, but you cannot do both.

And yet this dress code for young children, despite being so strict, is a relatively recent phenomenon. Until the end of the nineteenth century, even five-year-old children were being dressed in more-or-less unisex white dresses, according to sociologist Jo

Paoletti. The introduction of coloured fabrics for young children's clothing marked the beginning of the move towards our current pink-blue labelling of gender, but it took nearly half a century for the rules to settle into place. For a time, pink was preferred for boys, because it was 'a decided and stronger' colour, a close relative to red, symbolising 'zeal and courage'. Blue, being 'more delicate and dainty' and 'symbolic of faith and constancy' was reserved for girls. Only towards the middle of the twentieth century did existing practices become fixed.[1]

Yet so thoroughly have these preferences become ingrained that psychologists and journalists now speculate on the genetic and evolutionary origins of gendered colour preferences that are little more than fifty years old.[2] For example, a few years ago an article in an Australian newspaper discussed the origins of the pink princess phenomenon. After trotting out the ubiquitous anecdote about the mother who tried and failed to steer her young daughter away from the pink universe, the journalist writes that the mother's failure 'suggests her daughter was perhaps genetically wired that way' and asks, 'is there a pink princess gene that suddenly blossoms when little girls turn two?' Just in case we mistake for a joke the idea that evolution might have weeded out toddlers uninterested in tiaras and pink tulle, the journalist then turns to prominent child psychologist Dr. Michael Carr-Gregg for further insight into the biological basis of princess mania: 'The reason why girls like pink is that their brains are structured completely differently to boys', he sagely informs us. 'Part of the brain that processes emotion and part of the brain that processes language is one and the same in girls but is completely different in boys.' (Now where have we heard *that* before?) 'This explains so much – you can give a girl a truck and she'll cuddle it. You can give a boy a Barbie doll and he'll rip its head off.'

But what is also overlooked is *why*, according to Paoletti, children's fashions began to change. Dresses for boys older than two years old began to fall out of favour towards the end of the nineteenth century. This was not mere whim, but seemed to be in response to concerns that masculinity and femininity might not, after all, inevitably unfurl from deep biological roots. At the same time that girls were being extended more parental licence to be physically active, child psychologists were warning that 'gender distinctions *could be taught* and *must* be'. Some pants, please, for the boys. After the turn of the century, psychologists became more aware of just how sensitive even infants are to their environments. As a result, '[t]he same forces that had altered the clothing styles of preschoolers – anxiety about shifting gender roles and the emerging belief that gender could be taught – also transformed infantswear.'[4]

In other words, colour-coding for boys and girls once quite openly served the purpose of helping young children learn gender distinctions. Today, the original objective behind the convention has been forgotten. Yet it continues to accomplish exactly that, together with other habits we have that also draw children's attention to gender, as a number of developmental psychologists have insightfully argued.[5]

Imagine, for a moment, that we could tell at birth (or even before) whether a child was left-handed or right-handed. By convention, the parents of left-handed babies dress them in pink clothes, wrap them in pink blankets and decorate their rooms with pink hues. The left-handed baby's bottle, bibs and dummies – and later, cups, plates and utensils, lunch box and backpack – are often pink or purple with motifs such as butterflies, flowers and fairies. Parents tend to let the hair of left-handers grow long, and while it is still short in babyhood a barrette or bow (often pink) serves as a stand-in. Right-handed babies, by contrast, are never dressed in pink; nor do they ever have pink accessories or toys. Although blue is a popular colour for right-handed babies, as they get older any colour, excluding pink or purple, is acceptable. Clothing and other items for right-handed babies and children commonly portray vehicles, sporting equipment and space rockets; never butterflies, flowers or fairies. The hair of right-handers is usually kept short and is never prettified with accessories.

Nor do parents just segregate left- and right-handers symbolically, with colour and motif, in our imaginary world. They also distinguish between them verbally. 'Come on, left-handers!' cries out the mother of two left-handed children in the park. 'Time to go home.' Or they might say, 'Well, go and ask that right-hander if *you* can have a turn on the swing now.' At playgroup, children overhear comments like, 'Left-handers love drawing, don't they?', and 'Are you hoping for a right-hander this time?' to a pregnant mother. At preschool, the teacher greets them with a cheery, 'Good morning, left-handers and right-handers.' In the supermarket, a father says proudly in response to a polite enquiry, 'I've got three children altogether: one left-hander and two right-handers.'

And finally, although left-handers and right-handers happily live together in homes and communities, children can't help but notice that elsewhere they are often physically segregated. The people who care for them – primary caregivers, child care workers and kindergarten teachers, for example – are almost all left-handed, while building sites and garbage trucks are peopled by right-handers. Public toilets, sports teams, many adult friendships and even some schools, are segregated by handedness.

You get the idea.

It's not hard to imagine that, in such a society, even very young children would soon learn that there are two categories of people – right-handers and left-handers – and would quickly become proficient in using markers like clothing and hairstyle to distinguish between the two kinds of children and adults. But also, it seems more than likely that children would also come to think that there must be something fundamentally important about whether one is a right-hander or a left-hander, since so much fuss and emphasis is put on the distinction. Children will, one would imagine, want to know what it *means* to be someone of a particular handedness and to learn what sets apart a child of one handedness from those with a preference for the other hand.

We tag gender in exactly these ways, all of the time. Anyone who spends time around children will know how rare it is to come across a baby or child whose sex is not labelled by clothing, hairstyle or accessories. Anyone with ears can hear how adults constantly label gender with words: *he*, *she*, *man*, *woman*, *boy*, *girl* and so on. And we do this even when we don't have to. Mothers reading picture books, for instance, choose to refer to storybook characters by gender labels (like *woman*) twice as often as they choose nongendered alternatives (like *teacher* or *person*).[6] Just as if adults were always referring to people as left-handers or right-handers (or Anglos and Latinos, or Jews and Catholics), this also helps to draw attention to gender as an important way of dividing up the social world into categories.

This tagging of gender – especially different conventions for male and female dress, hairstyle, accessories and use of makeup – may well help children to learn how to divvy up the people around them by sex. We've seen that babies as young as three to four months old can discriminate between males and females. At just ten months old, babies have developed the ability to make mental notes regarding what goes along with being male or female: they will look longer, in surprise, at a picture of a man with an object that was previously only paired with women, and vice versa.[7] This means that children are well-placed, early on, to start learning the gender ropes. As they approach their second birthday, children are already starting to pick up the rudiments of gender stereotyping. There's some tentative evidence that they know for whom fire hats, dolls, makeup and so on are intended before their second birthday.[8] And at around this time, children start to use gender labels themselves and are able to say to which sex they themselves belong.[9]

It's at this critical point in their toddler years that children lose their status as objective observers. It is hard to merely dispassionately note what is for boys and what is for girls once you realise that you are a boy (or a girl) yourself. Once children have personally relevant boxes in which to file what they learn (labelled 'Me' versus 'Not Me'), this adds an extra oomph to the drive to solve the mysteries of gender.[10] Developmental psychologists Carol Martin and Diane Ruble suggest that children become 'gender detectives', in search of clues as to the implications of belonging to the male or female tribe.[11] Nor do they wait for formal instruction. The academic literature is scattered with anecdotal reports of preschoolers' amusingly flawed scientific accounts of gender difference:

> [O]ne child believed that men drank tea and women drank coffee, because that was the way it was in his house. He was thus perplexed when a male visitor requested coffee. Another child, dangling his legs with his father in a very cold lake, announced 'only boys like cold water, right Dad?' Such examples suggest that children are actively seeking and 'chewing' on information about gender, rather than passively absorbing it from the environment.[12]

In fact, young children are so eager to carve up the world into what is female and what is male that Martin and Ruble have reported finding it difficult to create stimuli for their studies that children see as gender neutral, 'because children appear to seize on any element that may implicate a gender norm so that they may categorize it as male or female.'[13] For instance, when creating characters from outer space for children, it proved difficult to find colours and shapes that didn't signify gender. Even something as subtle as the shape of the head could indicate gender in the eyes of the children: aliens with triangular heads, for example, were seen as male.[14] (Later, we'll see why.) And experimental studies bear out children's propensity to jump to *Men Are from Mars*, *Women Are from Venus*–style conclusions on rather flimsy evidence. Asked to rate the appeal of a gender-neutral toy (which girls and boys on average like the same amount), boys assume that only other boys will like what they themselves like; ditto for girls.[15]

It's hardly surprising that children take on the unofficial occupation of gender detective. They are born into a world in which gender is continually emphasised through conventions of dress, appearance, language, colour, segregation and symbols. Everything around the child indicates that whether one is male or female is a matter of great importance. At the same time, as we'll see in the next chapter, the information we provide to children, through our social structure and media, about what gender means – what goes with being male or female – still follows fairly old-fashioned guidelines.

orty years ago, psychologists Sandra and Daryl Bem decided to raise their young children Jeremy and Emily in a gender-neutral way. Their goal was to restrict as much as they could their young children's knowledge of the 'cultural correlates' of gender, at least until they were old enough to be critical of stereotypes and sexism.

What, exactly, did this involve?

Theirs was a two-pronged strategy. First, the Bems did all that they could to reduce the normally ubiquitous gender associations in their children's environment: the information that lets children know what toys, behaviours, skills, personality traits, occupations, hobbies, responsibilities, clothing, hairstyles, accessories, colours, shapes, emotions and so on go with being male and female. This entailed, at its foundation, a meticulously managed commitment to equally shared parenting and household responsibilities. Trucks and dolls, needless to say, were offered with equal enthusiasm to both children; but also pink and blue clothing, and male and female playmates. Care was taken to make sure that the children saw men and women doing cross-gender jobs. By way of censor-ship, and the judicious use of editing, WhiteOut and marker pens, the Bems also ensured that the children's bookshelves offered an egalitarian picture-book world:

[M]y husband and I got into the habit of doctoring books whenever possible so as to remove all sex-linked correlations.

We did this, among other ways, by changing the sex of the main character; by drawing longer hair and the outline of breasts onto illustrations of previously male truck drivers, physicians, pilots, and the like; and by deleting or altering sections of the text that described females or males in a sex-stereotyped manner. When reading children's pictures books aloud, we also chose pronouns that avoided the ubiquitous implication that all characters without dresses or pink bows must necessarily be male: 'And what is this little piggy doing? Why, he or she seems to be building a bridge.'[1]

The second part of the Bems' strategy was to, in place of the usual information about what it means to be male or female, promote the idea that the difference between males and females lies in their anatomy and reproductive functions. Your typical preschooler enjoys a detailed knowledge of gender roles, but remains a bit hazy regarding the hard, biological fact that males differ from females when it comes to the allocation of such items as penises, testicles and vaginas.[2]

Not so, for the Bem children:

[O]ur son Jeremy, then age four, . . . decided to wear barrettes [hair slides] to nursery school. Several times that day, another little boy told Jeremy that he, Jeremy, must be a girl because 'only girls wear barrettes.' After trying to explain to this child that 'wearing barrettes doesn't matter' and that 'being a boy means having a penis and testicles,' Jeremy finally pulled down his pants as a way of making his point more convincingly. The other child was not impressed. He simply said, 'Everybody has a penis; only girls wear barrettes.'

Unlike their peers, Jeremy and Emily were discouraged from using socially determined trappings such as hairstyle, clothing, accessories or profession as a guide to a person's biological sex. If the children asked whether someone was male or female, their parents 'frequently denied certain knowledge of the person's sex, emphasizing that without being able to see whether there was a penis or a vagina under the person's clothes, [they] had no definitive information.'[3]

Step forward, please, all those parents who go to similar lengths to protect their children from acquiring prevailing cultural assumptions about gender. And do try to avoid being trampled in the rush.

The Bems' efforts, I think you'll agree, seriously outclass what we normally, generously, think of as gender-neutral parenting. They were, in Sandra Bem's own words, 'an unconventional family'.[4] Some readers will be cheering in admiration, while others roll their eyes with a quiet groan. But whatever your opinion of a parent who teases, 'What do you mean that you can tell Chris is a girl because Chris has long hair? Does Chris's hair have a vagina?'[5] we can all agree that the intensity and scope of the Bems' efforts offer a helpful hint as to just how gendered children's environments are. To this day social structure, media and peers offer no shortage of information to children about masculinity and femininity.

The gendered patterns of our lives can be so familiar that we no longer notice them, as this anecdote reported by legal scholar Deborah Rhode slyly makes plain:

> One mother who insisted on supplying her daughter with tools rather than dolls finally gave up when she discovered the child undressing a hammer and singing it to sleep. 'It must be hormonal,' was the mother's explanation. At least until someone asked who had been putting her daughter to bed.[6]

Yet children, with their fresher observational powers, take note. 'Russell is a funny Daddy', commented an astute three-year-old visitor to our home, observing our household's shared parenting practices. 'He stays at home like a Mummy.' Children dropping in to play after school sometimes turn to our son and ask in surprise, 'Why is your dad home?' (And more than one child of our acquaintance has disillusioned a boastful father with the information that, to the contrary, *Russell* is the best Daddy in the world.) Russell, my husband, is indeed 'funny' statistically speaking (as well as in other ways that need not concern us here). Whatever you think of the rights, wrongs or reasons for it, it is an empirical fact that children are born into an environment in which it is overwhelmingly women who service the child's – and family's – needs. Rare indeed are the children who see their father do more domestic labour than their mother. In fact, as we saw in Chapter 7, there seems to be *no* work arrangement between mothers and fathers – including his unemployment or her massive salary – that lets women off the domestic hook. Even the rare families who genuinely value each parent's career and leisure time equally, and fairly split the domestic load may find themselves dismissed as an aberrant (or 'funny') data point, as Australian psychologist Barbara David and colleagues have suggested. They note that in a classic study, children were shown a video of men and women playing a game, with the men performing one kind of ritual and the women another. Girls copied the women's ritual, and boys the men's, but only after they had confirmed for themselves that this is what women (or men) *in general* did, and not just one particular woman or man. 'Thus a parent,' suggests David, 'no matter how loving or loved, cannot be a model for appropriate gender behaviour, unless the child's exposure to the wider world (for example, through friendship groups and the media) suggests that the parent is a *representative* or prototypical male or female'.[7]

If so, the egalitarian parent can look forward to being undermined on a daily basis. For, as it happens, neither children, nor children's media, are renowned for their open-minded approach to gender roles.

Young children, for instance, certainly don't tend to take the expansive, laissez-faire approach when it comes to gender. Last year, when my son was in kindergarten, he asked a classmate if he could look at her book. 'No', the little girl told him. 'Boys aren't *allowed* to look at books about fairies.' The child well-versed in gender stereotypes is not shy about letting it be known that a peer has crossed the line. When developmental psychologists unobtrusively watch what goes on in preschool classrooms, they find that children receive distinctly cooler responses from peers when they play in gender-inappropriate ways. Developmental psychologist Beverly Fagot found that comments as blunt as 'you're silly, that's for girls' and 'that's dumb, boys don't play with dolls' were especially reserved for boys.[8] But boys and girls alike are treated to little pointers when other children praise, imitate and join in certain types of play, but criticise, disrupt or abandon other activities. Unsurprisingly, this peer feedback seems to influence children's behaviour, making it more stereotypical.[9] Peers' responses appear to act as reminders to children that their behaviour doesn't follow gender rules, because they are particularly effective in bringing cross-gender behaviour to an end. In fact, it seems as though even the *prospect* of 'jeer pressure' may change young children's behaviour. Preschool children spend more time playing with gender-appropriate toys when an opposite-sex peer is nearby, in comparison with play in the absence of another child.[10] Likewise, four- to six-year-old boys express more interest in playing with boyish toys when they are with peers than when they are on their own.[11] The sensitivity of preschool boys to breaking unwritten gender rules was very much in evidence in a group of preschool children in the UK, who were observed by David Woodward. Younger boys who generally would not play with dolls at preschool (one boy is described furtively dressing and undressing a doll under the table, looking over his shoulder all the while to be sure he wasn't spotted by other boys) would nonetheless happily play with them at home. And once a rather dominant and socially conservative group of boys left the preschool, the gender rules relaxed; more of the remaining boys started to play with dolls, and in the home corner.[12]

The media, like peers, also offer lessons in the cultural

correlates of gender. Rather than embrace the opportunity to present an imaginary world that offers children a glimpse of possibilities beyond the reality of male and female social roles, children's media often continue to constrict gender roles, sometimes even with more rigidity than does the real world:

> Meet the Jetsons, the family of the future, as imagined by cartoonists in the 1960s. George flies to work in his bubble car while Jane whips up instant meals from a tiny pill using a nuclear energy oven. Even though the Jetsons live in a bio-morphic building with a robot for a maid, in terms of gender relations, they might as well be the Flintstones. Dad works and worries about money while mom either stays at home or shops . . . Although the show's creators were highly imaginative when it came to the technological gadgets . . . they could not envision the real change that families underwent.[13]

In picture books of this time, too, it seemed to be easier for writers and illustrators to conceive of wonderful fantasy worlds and adventures than it was for them to imagine a woman in a paid occupation. A classic study published in 1972 analysed picture books awarded the prestigious Caldecott Medal; in particular, the eighteen winners and runners-up for this award between 1967 and 1971. The authors point out the absurdity of the fact that 40 percent of women (at that time) were in the labour force, and yet '*not one* woman in the Caldecott sample had a job or profession.'[14] Many classic picture books that children still enjoy were written during this period, in which the unwritten rule seems to be that a woman character should be illustrated wearing an apron, or not at all. And even today, contemporary research shows that picture-book women are still cracking their heads against the glass ceiling, venturing only rarely into traditionally male occupations, as well as being less likely to work outside the home than picture-book men.[15]

And why indeed should they, when the ensnaring of a rich and handsome prince can provide long-term financial security? *Disney Princess* magazine, targeted at the sophisticated two- to four-year-old-girl market, is just one manifestation of the now fantastically successful pink princess phenomenon. The princess genre offers lessons in how to achieve what old-school feminists refer to in tight-lipped fashion as the traditional feminine ideal, that is, how to be pretty, caring and catch a husband. No pursuit, it seems, is too trivial for (some, at least) modern-day princess books and magazines: little princesses are advised to '[a]ccessorise to impress' and, in order that their hair might look as pretty as Belle's when she danced with the Beast, to 'try a deep conditioner'.[16] Once the preschooler becomes too worldly for innocent fairy-tale fashion, romance and marriage, she can graduate at age five to more grown-up versions of the same focus on beauty and romance, thanks to magazines like *Barbie Magazine*, three-quarters of the content of which is devoted to (in order of greatest to least prevalence) crushes, celebrities, fashion and beauty.[17]

But even in higher-quality children's literature, more subtle stereotypes remain. Diane Turner-Bowker examined how males and females were described in the forty-one Caldecott winners and runners-up from 1984 to 1994. One gender was most commonly described as, among other adjectives, *beautiful, frightened, worthy, sweet, weak* and *scared* in the stories; the other gender as *big, horrible, fierce, great, terrible, furious, brave* and *proud*. (If you're not sure which sex is being described in these two lists, ask your nearest gender-neutrally reared preschooler; he or she will be sure to know.) Unsurprisingly, the adjectives for males were rated as more powerful, active and masculine than those used for females.[18] And we all know which type of person we'd rather have with us on an adventure. '[G]irls are often left out of the adventure, the thrill, the plot, the *picture*' even today in the Caldecott award winners, point out *Packaging Girlhood* authors Sharon Lamb and Lyn Brown, who combed through them all in search of a female adventuress. 'By the time you get to *Mirette on the High Wire*, the only book in the past twenty years that features a girl in an adventure, you know this isn't coincidence.'[19] (Sadly, even poor Mirette is soon misremembered as being stereotypically feminine rather than the 'gallant, resourceful little girl' she really is.)[20]

Even so, it is easier to find an adventurous girl than a sissy boy. The bucking of gender stereotypes in young children's books is a task usually performed by female characters, many researchers have found. Just as in the real world women have been quicker to forge forth into the masculine world of work than men have been to sink back into domesticity, in children's books, too, it is mostly females who do the crossing of gender boundaries. Amanda Diekman and Sarah Murnen, for example, compared twenty popular and enduring books for elementary school children, half of which enjoyed the recommendation of being nonsexist by educational commentators (like *Alice in Wonderland* and *Harriet the Spy*), while the remainder had been classified as sexist (such as *Charlie and the Chocolate Factory* and *The Wheel on the School*). They found that it was the taking up of masculine traits, roles and leisure activities by female characters that set apart the supposedly nonsexist books from the sexist ones. Yet these nonsexist books were no more likely than the sexist ones to portray males as femininely tender and compassionate, in domestic servitude or contentedly engaged with girlish activities or toys.[21]

Reviews of elementary school readers (books used to teach reading) in the United States similarly conclude, 'No sissy boys here'.[22] And there are not too many sissy fictional fathers, either. Among Caldecott books from 1995 to 2001 and best-selling children's books of around the same time, fathers are not only scarce, but also lacking in good cot-side manner, being 'presented as unaffectionate and as indolent in terms of feeding, carrying babies, and talking with children.'[23] Children's TV programmes still often rely on gender stereotypes, even in children's educational programming.[24] Dora the Explorer – the intrepid Latina adventuress – is a notable exception. (Check out the Dora merchandise on the Fisher-Price Web site, however, and you will quickly uncover the familiar themes of princesses, mermaids and fashion.) And of course toy advertisements make it quite clear for whom – boys or girls – particular toys and activities are intended. Lamb and Brown watched hours of Nickelodeon, taking note of the advertisements in between popular programmes. On a typical day, they saw boys playing with Legos, cars and action figures, and girls playing with princesses, fairies, kitchen sets and fashionably dressed and accessorised dolls.[25] And children take note of who is playing with what: when researchers doctored a commercial for a Playmobil Airport Set to show girls, as well as boys, playing with the toy, first- and second-grade children shown this altered commercial were nearly twice as likely to think that the toy was for girls as well as boys, compared with children who saw the commercial in its traditional, boys-only form.[26]

Media also distinguish between males and females in a more subtle way: importance. 'Children scanning the list of titles of what have been designated as the very best children's books are bound to receive the impression that girls are not very important because no one has bothered to write books about them. The content of the books rarely dispels this impression', remarked Lenore Weitzman and colleagues in their classic review of Caldecott winners,[27] nearly a third of which had no female characters at all. And of course there are characters, and then there are *main* characters. The Dr. Seuss books are rightly classics, adored by children and a joy of rediscovery for parents. Yet as Lamb and Brown observe, in all the forty-two books he wrote, not one has a female lead in its central story.[28] The power of the media to dish up a stripped-down, concentrated version of cultural values enables it to represent the higher status of males in this uncomfortably blunt fashion. Even in contemporary picture books, researchers find that this is a habit that dies hard, with writers and illustrators still less inclined to feature female characters. For example, the most recent analysis of the Caldecott winners and runners-up, together with 155 best-selling children's books around the same time, found that males, overall, were featured nearly twice as often as females in title roles, and they appeared in about 50 percent more pictures.[29]

Nor does the use of gender-ambiguous animals or characters in books help to increase female numbers. This is because mothers almost always label gender-neutral characters in picture books as male.[30] If it doesn't look like a female, it's male. I've tried labelling neutral animals and characters as female when reading to my children – it feels extremely unnatural, as you will discover if you try for yourself. (The reason is probably that we have a tendency to think of people or creatures as male unless otherwise indicated. In other words, as has been long observed, men are people, but women are women.) As within the pages of books, females tend to be underrepresented on TV and computer screens, and to miss out on central roles in advertisements and even cereal boxes.[31] A recent survey of 19,664 children's programmes in twenty-four countries found that only 32 percent of main characters are female.[32] (This drops to an even more dismal 13 percent when it comes to nonhuman creatures like animals, monsters and robots.) And, a survey of the 101 top-grossing G-rated movies from 1990 to 2005 found that less than a third of the speaking roles go to females, with no signs of improvement over time.[33] As the Web site of the Geena Davis Institute, which sponsored the research, asks, 'What message does this send to young children?'[34]

With fervent and tireless testing of hypotheses taking place – and with such a wealth of data to work with – it's hardly surprising that by the time they are four years old children are already remarkably advanced gender theorists. (One can even, at a stretch, imagine a panel of preschoolers coming up with, or perhaps even improving upon, certain popular book titles such as: *Men Are Like Waffles, Women Are Like Spaghetti*; *Why Men Don't Iron*; and *Why Men Don't Have a Clue and Women Always Need More Shoes*.) To the preschooler, information about which gender goes with hammers and fire hats, and brooms and baby bottles, was covered way back in Gender Stereotyping 101.[35] They know it all. But what is perhaps most amazing is that, without even troubling to read the latest best-selling exposition of biological essentialism, they are using this database of cultural correlates to draw out some general, abstract principles. Social psychologists Laurie Rudman and Peter Glick pithily characterise the content of gender stereotypes as 'bad but bold' (with males being tough, competitive and assertive) versus 'wonderful but weak' (with females stereotyped as being gentle, kind and soft).[36] And preschoolers, it seems, are already working this out for themselves. 'Few men keep bears', as developmental psychologist Beverly Fagot and colleagues pointed out. And yet four-year-olds reliably classify a fierce looking bear as for boys. They can even classify different shapes, textures and emotions (like angular, rough and anger) as male and female.[37] This is why the triangle-headed creatures from outer space mentioned earlier were categorised as male – all those angles. Indeed, so powerful are these metaphorical gender cues that five-year-old children will confidently declare that a spiky brown tea set and an angry-looking baby doll dressed in rough black clothing are for boys, while a smiling yellow truck adorned with hearts and a yellow hammer strewn with ribbons are for girls.[38]

This is truly remarkable, when you think about it. Heaven knows, I've heard enough parents openly labelling certain sports, toys, activities, behaviours and personality traits as being for boys or for girls. In one month alone, I heard people referring to colouring in a dinosaur, playing soccer, being noisy and wanting to press elevator buttons as boy things. But you don't often hear a parent exclaiming, 'No, *no*, Jane! Angles are for boys, not girls. Take the curved one.' Yet even before they reach school, children can go well beyond the surface of gender associations and make inferences about nothing less than male and female inner nature itself. They also seem to learn, uncomfortably young, that females are 'other'. When Barbara David asked four- and five-year-old children to choose items that would show a martian what human beings were like, the girls chose a mix of female and male objects (such as guns and dolls), whereas the boys chose almost only male items.[39]

All of this was what the Bems were trying to avoid. As we imagine them bent over their children's picture books, carefully whiting out beards and drawing in breasts, we can see why, without a doubt, they would not be terribly impressed by the despairing tales of parents who simply offer their children a few nontraditional toys.

A few years ago, when the Australian feminist writer Monica Dux wrote an opinion piece criticising parents' tolerance for the pink princess phenomenon, one angry respondent presented her own disapproval as evidence that her daughter's passion for pink was a manifestation of her true self that it would be somehow wrong to deny:

> On giving birth to a daughter, I swore that she wouldn't be smothered in frilly pink clothes, and that she would play with cars and with stuffed animals. As it turns out, my child is a person in her own right. She loves all things pink and frilly. . . . I worry . . . that if I deny her this pleasure, then it is just the beginning of a long road where I tell her that she is not allowed to be herself but rather that she must become what I want her to be.[1]

Fine for millions of marketing dollars to be spent promoting a pink, frilly world to girls. Parents, however, should keep their opinions to themselves lest they unduly influence children's preferences! But also, because gendered preferences often appear to develop despite their best efforts, parents often assume that they must come from within the child: the biology-as-fallback position described by Emily Kane. Yet as New York University developmental psychologist Diane Ruble points out, '[i]t requires little detective work for children to notice some of the most blatant physical characteristics associated with females: pink, frilly, and dresses.'[2] She, Cindy Miller, and colleagues asked preschoolers the open-ended question, 'Tell me what you know about girls. Describe them.' This way, they could see what it was about girls that came most quickly and easily to children's minds. The most frequent answer related to appearance: girls have long hair, girls are pretty, girls wear dresses – that kind of thing.[3] (Feminine Beauty Ideal: 1. Old-fashioned feminism: 0.) By contrast, the preschoolers' descriptions of boys centred more on the sorts of activities that boys do and their rough, active, personality traits.

How does this kind of knowledge, amassed from an early age, influence children? As we've seen, children are born into a world in which gender is continually emphasised through conventions of dress, appearance, language, colour, segregation and symbols. Everything around the child indicates that whether one is male or female is a matter of great importance. Meanwhile, at about two years of age, children discover on which side of the divide they are located. It remains to be seen, in my view, whether subtle gender differences in babies' toy preferences before they know their own sex can be explained by socialisation by parents, unwitting or otherwise. But once children know their own sex, in theory they can start to take socialisation into their own hands.

And it's plausible to think that they will. Gaining membership to a group, any group, normally brings a money-back guarantee of favouritism. In the infamous minimal group studies conducted by Henri Tajfel and colleagues, adults are randomly assigned to completely trivial groups. For example, they are asked to estimate the number of dots in an array, and then categorised as either a dot overestimator or a dot underestimator. It's hard to imagine a categorisation of less psychological significance. And yet membership of even such arbitrarily assigned and short-lived social categories can engender a warm glow towards fellow dot overestimators (or underestimators) that does not extend so far as those who take a different approach to dot guesstimating.[4]

Children, it turns out, are also susceptible to an in-group bias to prefer what belongs to their group. Recent work by Rebecca Bigler and colleagues has shown that this is especially the case when groups are made visually distinct, and authority figures use and label the groups. In one study, three- to five-year-old pre-schoolers in two child-care classrooms were randomly assigned to the Blue group or the Red group. Over a three-week period all the children wore a red or blue T-shirt every day (according to the group to which they'd been assigned). In one classroom, the teachers left it at that. The colour groups were not mentioned again. But in the other classroom, the teachers made constant use of the two categories. Children's cubbies were decorated with blue and red labels, at the door they were told to line up with Blues on this side and Reds on that side, and they were regularly referred to by group label ('Good morning, Blues and Reds'). At the end of the three weeks, the experimenters canvassed each child's opinion on a number of matters. They found that being categorised as a Red or a Blue for just three weeks was enough to bias children's views. The children, for example, preferred toys they were told were liked by their own group and expressed a greater desire to play with other Red (or Blue) children. While some forms of favouritism were common to all the children, more was seen in kids from the classroom in which teachers had made a bigger deal out of the Red versus Blue dichotomy.[5]

Just imagine how powerfully exactly the same psychological mechanisms can drive in-group pride and out-group prejudice when it comes to gender. In the young child's world, gender is the social category that stands out above all others, right from the start. Conventions of clothing and accessories mean that gender is extremely obvious visually, and boys and girls may be regularly labelled and organised ('Now it's the boys' turn to wash their hands') by gender, especially in early education settings.[6] And, unlike adults and older children, younger children don't tend to have other social categories like *jock*, *doctor*, *Christian* or *artist* with which to identify.[7] The drive for group belonging may explain why young children insist on girlish or boyish behaviour or dress even in the face of parental displeasure, suggest Diane Ruble and colleagues.[8]

So for the self-socialising preschool girl, a puff of pink frills lends solidity to an important group identity based on gender. Every semester, my youngest son's kindergarten has a dress-up day. One little girl in a cat costume walked into the room to discover that every other girl, without exception, was dressed up as either a princess or a fairy. She burst into tears and wailed to her mother, 'I should have worn my princess dress!' On the next dress-up day, she did.

Likewise, we can expect boys to be drawn to toys or activities that fit with their sophisticated, metaphorical understanding that 'tough' is for boys:

> In one study, researchers transformed a pastel 'My Little Pony' by shaving the mane (a soft 'girlish' feature), painting it black (a 'tough' colour), and adding spiky teeth (for an aggressive demeanour). Both boys and girls classified the altered pony as a boy's toy, and most of the boys (but not the girls) were extremely interested in obtaining one.[9]

The five-year-old girls in this study, by the way, 'were enchanted by . . . the lavender-satin-covered guns and holster, and the pink-furred war helmet'.[10]

A child's toy preferences are no doubt influenced by a whole host of factors, with his or her gender knowledge being just one part of a complicated mix. But nonetheless, although this literature is somewhat mixed, overall it does suggest that gender identity (*I am a boy*) and gender stereotype knowledge (*Boys don't play with this toy*) motivate gender stereotypical play.[11] For example, psychologist Kristina Zosuls and her colleagues recently tracked what seemed to be the very start of this process in children who were not yet two years old. They looked at toddlers' play behaviour at both seventeen and twenty-one months of age, to see how it changed as the children started to use gender labels (like *lady* and *boy*) to refer to themselves or others. At seventeen months, boys and girls were equally interested in the doll, tea set, brush and comb set and blocks, although girls spent less time playing with the truck. But four months later, girls had increased their doll play and boys had decreased it. A closer look at this shift revealed that gender labelling was associated with more gender-stereotypical play.[12]

With older children, who are in no doubt about their gender identity, you can manipulate gender labels and watch what happens. In school-aged children, subtle gender labels like 'This is a test to see how good you would be at mechanics or at operating machinery' (versus needlework, sewing or knitting) affect children's performance in stereotype-consistent ways.[13] And with children under age six, putting a gender label on a gender-neutral toy is a reliable way of creating gender-stereotypical behaviour. For example, four-year-old children will play for three times as long with a xylophone or balloon if it is labelled as being for their own sex rather than for children of the other sex. A less attractive gender-neutral toy can be rendered instantly more desirable simply by applying the correct gender-label. And conversely, an attractive novel toy becomes less so when labelled as for the other sex.[14]

It's also possible to make even decidedly gender-stereotyped toys more appealing, especially perhaps to girls, by showing them that they can be played with by the other sex, too. In one small study, Rebecca Bigler and her colleagues identified eight preschoolers, four girls and four boys, who reliably avoided toys traditionally played with by the other sex. These children were then read two carefully constructed tales that unsubtly exploded gender stereotypes at every turn: one story starred the exuberant Sally Slapcabbage and her pilot mother; the second featured Billy Bunter, who finds and cherishes a talking doll. Thanks to the stories, two of the four boys overcame a little of their reluctance to explore their feminine side on the playmat, venturing to play with the sorts of toys they would normally ignore. Yet even more remarkable was the effect of the stories on three of the four girls. After just a few readings of the counterstereotypic stories, these girls abandoned stroller, baby doll and ironing board to experiment with fire trucks, blocks and helicopters. By the last few days of the experiment these girls were playing almost exclusively with the boyish toys.[15] After just a few doses of Sally Slapcabbage, one would be hard-pressed to distinguish these once ultrafeminine preschoolers from the girls with congenital adrenal hyperplasia (exposed to unusually high levels of foetal testosterone) we met in Chapter 11.

So what are we to make now of the little girl tucking 'baby truckie' into bed? If we focus in just on her, then yes, the failure of gender-neutral parenting to achieve its aim will indeed seem comical. But widen your field of vision to include the less-visible cultural waters in which the sponges that are our children are immersed, and the real joke is the idea that children are being reared in a gender-neutral fashion. Emily Kane suggests that the rapidity with which highly educated and privileged parents fall back on biological explanations reflects their position at 'the vanguard of a limited sociological imagination'.[16] Harsh but, I think, fair.

Children's views about gender differences reach 'peak rigidity' between five and seven years of age.[17] From then on, they increasingly understand that it is not only boys who like to be active, and make things and sometimes be nasty, and it is not only women who can be affectionate, cry, and clean and tidy the house. (The few children who don't come around to this insight often go on to have very successful careers writing popular books based on rigid gender stereotypes.)[18] But even as their growing cognitive flexibility enables them to consciously modify or even reject certain gender stereotypes, we can only presume that these stereotypical gender associations linger on, continuing to be reinforced by the patterns of a half-changed world. There they will be, ready to flesh out the details of the self-concept whenever the social context brings a gender identity to the fore. There they will be as they judge their work colleagues and negotiate privileges and patterns in their romantic relationships. There, perhaps, they will be as they interpret sex differences in the brain. And there they will be if they become parents themselves.

And so it goes on.

EPILOGUE:
AND S-T-R-E-T-C-H!

When a distinguished man of Harvard makes a few ill-advisedly public comments about women's limited aptitude for a male-dominated profession, you can be sure there will be controversy. So discovered Professor Richard Cabot of Harvard Medical School, who in 1915 addressed the graduating class of the Woman's Medical College of Philadelphia. According to newspaper reports, Cabot suggested to these ambitious young women that female physicians are temperamentally and physically ill-suited to the more demanding branches of medicine. They should therefore, in his opinion, avoid general practice and research and instead restrict themselves to social service work.[1] As one newspaper headlined the event: *Doctor Man Calls Doctor Woman Unfit*. In the debate that followed, Cabot was defended by another distinguished medical professional, Dr. Simon Baruch, who agreed that women's nature curbed their options within medicine, arguing that women doctors, while enjoying the 'truly feminine temperamental qualities that spring from the biological maternal source', at the same time lack 'originality, logic, initiative, courage, and other distinctly masculine qualities'. Naturally, then, the 'true woman' will enjoy her greatest achievements in 'her own sphere' of 'nurturing civilisation'.

Dr. Baruch concluded his letter with the general concern that 'the dear women are "obsessed" with their fitness for all things masculine which blinds them to a sane view of their biological limitations.' He added, lest this remark be churlishly taken the wrong way, that '[t]hese lines are written in no spirit of controversy, simply to point out the irrevocable law of nature'.[2] By way of support, he referred to arguments made by the neurologist Dr. Charles L. Dana who, you will recall, was anxious that the upper half of the female spinal cord was a little on the light side for politics. And that is not all. Noting that 'women are rather more subject than men to the pure psychoses', Dana dolefully predicted that '[i]f women achieve the feministic ideal and live as men do, they would incur the risk of 25 per cent more insanity than they have now.'[3]

These fears do not look reasonable in the sharp focus of hindsight. At a time when, in the United States, women physicians in training outnumber men in dermatology, family medicine, psychiatry, paediatrics, OB/GYN, and are 'closing in fast' in internal medicine,[4] we can't help but judge a little harshly the career advice that women physicians should limit themselves to social welfare work. Dr. Cabot's prophecy that women physicians who ignore this advice are destined to become 'disappointed and dissatisfied' seems unnecessarily gloomy.[5] Likewise, Dr. Dana's worry that 'woman suffrage would . . . add to our voting and administrative forces the biological element of an unstable preciosity which might do injury to itself without promoting the community's good' appears to have been unfounded.[6] So far as I know, science has not documented any dangerous unravelling of feminine refinement and mental stability wrought by the sheer vulgarity of marking an X on a ballot. But we should not be too critical. These educated, intelligent men were simply worried by the prospect of social change. What would be the consequences for women who abandoned the nurturing roles for which they were biologically designed? Was it wise for them to be encouraged by feminists to seek access to the public spheres of men when they so clearly lacked the necessary mental and physical fitness? Had the biological limits of equality been reached, or even surpassed?

The error of these gloomy soothsayers, it's easy enough to see now, lay in their failure to adequately stretch the sociological imagination. So focused were they on locating the cause of inequality in some internal limitation of women – the lightweight brains, the energy-sapping ovaries, the special nurturing skills that leave no room for masculine ones – that they failed to see the injustice, as Stephen J. Gould put it, of 'a limit imposed from without, but falsely identified as lying within'.[7]

It would be better not to continue making the same mistake.

Take a look around. The gender inequality that you see is *in* your mind. So are the cultural beliefs about gender that are so familiar to us all. They are in that messy tangle of mental associations that interact with the social context. Out of this interaction emerges your self-perception, your interests, your values, your behaviour, even your abilities. Gender can become salient in the environment in so many ways: an imbalance of the sexes in a group, an advertisement, a comment by a colleague, a query about sex on a form, perhaps also a pronoun, the sign on a toilet door, the feel of a skirt, the awareness of one's own body. When the context activates gendered associations, that tangle serves as a barrier to nonstereotypical self-perception, concerns, emotions, sense of belonging and behaviour – and more readily allows what is traditionally expected of the sexes.

The fluidity of the self and the mind is impressive and is in continual cahoots with the environment. When social psychologists discover, for example, that mere words (like *competition*), everyday objects (like briefcases and boardroom tables), people or even scenery can trigger particular motives in us, or that similar role models can seep into our most private ambitions, it makes sense to start questioning the direction of causality between gender difference and gender inequality.[8] We are justified in wondering whether, as gender scholar Michael Kimmel suggests, 'gender difference is the product of gender inequality, and not the other way around.'[9]

Nor is gender inequality just part of our minds – it is also an inextricable part of our biology. We tend to think of the chain of command passing from genes, to hormones, to brain, to environment. (As biologist Robert Sapolsky describes this common misconception, 'DNA is the commander, the epicenter from which biology emanates. Nobody tells a gene what to do; it's always the other way around.')[10] Yet most developmental scientists will tell you that one-way arrows of causality are just *so* last century. The circuits of the brain are quite literally a product of your physical, social and cultural environment, as well as your behaviour and thoughts. What we experience and do creates neural activity that can alter the brain, either directly or through changes in gene expression. This neuroplasticity means that, as Kaiser puts it, the social phenomenon of gender 'comes into the brain' and 'becomes part of our cerebral biology'.[11]

As for hormones that act on the brain, if you cuddle a baby, get a promotion, see billboard after billboard of near-naked women or hear a gender stereotype that places one sex at a higher status than the other, don't expect your hormonal state to remain impervious. It won't. 'Even how we behave or what we think about can affect the levels of our sex hormones', point out *Gene Worship* authors Gisela Kaplan and Lesley Rogers.[12] This continuous interplay between the biological and the social means that, as Anne Fausto-Sterling has put it, 'components of our political, social, and moral struggles become, quite literally, embodied, incorporated into our very physiological being.'[13]

And so, when researchers look for sex differences in the brain or the mind, they are hunting a moving target. Both are in continuous interaction with the social context. Some researchers have even started to investigate how the brain, or hormones, respond differently while doing stereotyped tasks, depending on whether gender stereotypes are made salient.[14] And gender differences in the mind can shift from moment to moment: for example, as stereotype threat is created or dispersed, or self-identity changes. But also, our actions and attitudes change the very cultural patterns that interact with the minds of others to coproduce *their* actions and attitudes that, in turn, become part of the cultural milieu: in short, 'culture and psyche make each other up.'[15] When a woman persists with a high-level maths course or runs as a presidential candidate, or a father leaves work early to pick up the children from school, they are altering, little by little, the implicit patterns of the minds around them. As society slowly changes, so too do the differences between male and female selves, abilities, emotions, values, interests, hormones and brains – because each is inextricably intimate with the social context in which it develops and functions.

Where the convergence between female and male lives might end is anybody's guess. (A tip: the mistake is usually to undershoot.) But it *is* remarkable how similar the two sexes become, psychologically, when gender fades into the background. 'Love, tenderness, nurturance; competence, ambition, assertion – these are *human* qualities, and all human beings – both women and men – should have equal access to them', argues Kimmel.[16] Doesn't that sound nice? But it is still the case today that gender inequalities, and the gender stereotypes they evoke, interact with our minds in ways that create inequality of access.

Meanwhile, neuroscience is used by some in a way that it has often been used in the past: to reinforce, with all the authority of science, old-fashioned stereotypes and roles. 'The brain has frequently been the battle site in controversies over sex or race differences', as Ruth Bleier has noted.[17] Researching popular claims about the differences between male and female brains is not an activity that is particularly good for the blood pressure. The sheer audacity of the overinterpretations and misinformation is startling. Some commentators declare themselves to be courageous taboobreakers, who shout the scientific truth about sex differences into the hushed silence demanded by political correctness. But this is exactly how they shouldn't be regarded. For one thing, neurosexism is so popular, so mainstream, that I think it is difficult to argue that our attitude towards the supposedly unmentionable idea of innate sex differences is usually anything other than casual and forgiving. Can you imagine schools implementing brain-based single-race classrooms after seeing a few slides and pseudo-scientific facts about differences between 'black' brains and 'white' brains? If to talk about innate psychological differences between males and females was truly shocking and provocative, would publishers wave on to their hot list, or editors into their columns, books and articles that so misinform and mislead?

But also, to those interested in gender equality there is nothing at all frightening about good science. It is only carelessly done science, or poorly interpreted science, or the neurosexism it feeds, that creates cause for concern. Unfortunately, pointing out the problems can easily be framed as desperate nitpicking or the shooting of the messenger. Yet as Kaplan and Rogers point out, '[s]cepticism and rigorous science are not bad faults compared to moving prematurely to conclusions, especially when they influence social attitudes.'[18] These social attitudes about gender are an important part of the culture in which our brains and minds develop.

And it is into this powerful, pervasive web of social attitudes that children are born, parented and develop. Gender associations are soon learned, a legacy to last a lifetime, ready to be primed by the social context. Given the continual emphasis on gender in the young child's life, together with a rich fodder of information about its cultural correlates, it is hardly surprising that gender-neutral parenting fails. As sociologist Bronwyn Davies explains the problem for children:

> Children cannot both be required to position themselves as identifiably male or female and at the same time be deprived of the means of signifying maleness and femaleness. Yet this is what the vast majority of non-sexist programmes have expected them to do.[19]

The relentless gendering of everything around the child – from clothes, shoes, bedding, lunch boxes, even giftwrap, as well as the wider world around – makes this an all-but-impossible task. One effect of what has been described as 'the pernicious pinkification of little girls'[20] must surely be that gender becomes salient – to both boys and girls – with every rustle of pink tulle or twinkle of pretty shoes. How should children ignore gender when they continually watch it, hear it, see it; are clothed in it, sleep in it, eat off it?

Our minds, society and neurosexism create difference. Together, they wire gender. But the wiring is soft, not hard. It is flexible, malleable and changeable. And, if we only believe this, it will continue to unravel.

ACKNOWLEDGEMENTS

The research described in this book spans many different academic fields, and I am very grateful for the extremely helpful feedback and encouragement I received from the many experts in those areas, who generously found time to read and comment on portions of the manuscript. My warmest thanks go to Rebecca Bigler, Suparna Choudhury, Isabelle Dussauge, Ione Fine, Kit Fine, William Ickes, Anelis Kaiser, Emily Kane, Simon Laham, Carol Martin, Cindy Miller, Kristen Pammer, Alice Silverberg and especially Frances Burton, Anne Fine, Ian Gold, Giordana Grossi, Christine Kenneally, Neil Levy and Lesley Rogers. This book has been greatly improved by the expertise of these readers, as well as the many more academics who responded so helpfully to my queries. Any errors or misinterpretations that remain are my own.

My thanks also go to Jeanette Kennett, Neil Levy and the Centre for Applied Philosophy & Public Ethics, University of Melbourne, for support during and prior to the writing of this book. I am also grateful to everyone who played a part in making it fit for publication. My agent, Barbara Lowenstein, had an essential role in helping me develop my ideas. I thank her for her assistance and support. I am also extremely grateful to Simon Flynn and his colleagues Najma Finlay, Andrew Furlow and Sarah Higgins at Icon Books. Erica Stern was my endlessly patient and helpful contact at W. W. Norton, and I am very grateful to her, Carol Rose and my editor, Angela von der Lippe, for their many valuable comments and improvements to the manuscript. I also thank Laura Romain for her assistance.

Finally, my sincere thanks and gratitude go to my husband, Russell.

AUTHOR'S NOTE

It is, I imagine, extremely hard to say anything original about gender, and this has not been my goal. In synthesising material from many different disciplines my aim has been not to stand on the shoulders of others, but to report the view from that position in an accessible way. I am very appreciative of the important research, all done by others, cited in the long list of notes that follow. A few books stand out as deserving particular mention because of the important role they played in my own understanding of the areas they discuss, an influence that is hard to footnote in a book like this. When I first had the idea for this book, my concern about neuroscientific explanations of gender difference was limited to the crass popular interpretations of this literature. However, five books in particular laid the foundation for my understanding of the need for critical attention to the neuroscientific and neuro-endocrinological research itself. Ruth Bleier's *Science and Gender*, Anne Fausto-Sterling's two classics, *The Myths of Gender* and *Sexing the Body*, and Gisela Kaplan and Lesley Rogers's *Gene Worship* were eye-opening to me in their challenges and critiques of the unintended biases and unexamined assumptions often built into gender-difference research. Unexpectedly, *Sexual Science*, Cynthia Russett's historical account of Victorian sexual science, was also very helpful in this regard. Laurie Rudman and Peter Glick's recent book *The Social Psychology of Gender*, which comprehensively reviews this rapidly expanding field in a wonderfully coherent way, was an excellent resource. And a number of review articles and chapters by developmental psychologists Rebecca Bigler, Lynn Liben, Carol Martin, Cindy Miller, Diane Ruble and their colleagues were also extremely helpful. I am very grateful to all these scholars (and many more besides) for their work.

NOTES

INTRODUCTION

1 (Brizendine, 2007), pp. 166, 40, and 162, respectively.

2 (Brizendine, 2007), pp. 159 and 160, respectively.

3 http://www.gurianinstitute.com/meet_michael.php. Accessed December 2, 2008.

4 (Gurian, 2004), pp. 4 and 5, and p. 5.

5 (Sax, 2006), blurb.

6 (Gurian Institute, Bering, & Goldberg, 2009), p. 4.

7 (Gurian, Henley, & Trueman, 2001), p. 4.

8 (Gurian & Annis, 2008), jacket blurb.

9 (Gisborne, 1797), p. 21.

10 (Gisborne, 1797), p. 22.

11 (Baron-Cohen, 2003), p. 1. Emphasis in original.

12 (Levy, 2004), p. 319.

13 (Baron-Cohen, 2003), p. 185.

14 (Levy, 2004), pp. 319 and 320.

15 Mary Astell, *The Christian Religion* (1705). Quoted in (Broad, 2002), p. vii.

16 According to (Dorr, 1915).

17 Anon., 'Biology and Women's Rights', repr. *Popular Science Monthly*, 14 (Dec. 1878). Quoted in (Trecker, 1974), p. 363.

18 (Kimmel, 2004), p. vii.

19 (Kane, 2006b).

20 (Pinker, 2008), p. 5.

21 (Pinker, 2008), p. 266.

22 (Moir & Jessel, 1989), p. 21.

23 (Brizendine, 2007), pp. 36 and 37.

24 (Moir & Jessel, 1989), p. 20.

25 (Belkin, 2003), para. 60.

26 Social psychologists have marshalled evidence that suggests that we have a system

justification motive, 'whereby people justify and rationalise the way things are, so that existing social arrangements are perceived as fair and legitimate, perhaps even natural and inevitable.' (Jost & Hunyady, 2002), p. 119.

27 (Broad & Green, 2009), p. viii.

28 (Drake, 1696), p. 20. I'm grateful to Jacqueline Broad for bringing this quotation to my attention.

29 (Smith, 1998), p. 159.

30 E. L. Thorndike, 'Sex in Education', *The Bookman*, XXIII, 213. Quoted in (Hollingworth, 1914), p. 511.

31 (Mill, 1869/1988), p. 22.

32 Cora Castle, 'A statistical study of eminent women', *Columbia University contributions in philosophy and psychology*, vol. 22, no. 27 (New York: Columbia University, 1913), pp. vii, 1–90. Quoted in (Shields, 1982), p. 780.

33 (Malebranche, 1997), p. 130. I'm grateful to Jacqueline Broad for alerting me to this hypothesis.

34 See (Russett, 1989).

35 A phrase that originated with (Romanes, 1887/1987), p. 23. See (Russett, 1989), p. 36.

36 (Russett, 1989), p. 37.

37 See discussion in (Kane, 2006).

38 (Kitayama & Cohen, 2007), p. xiii.

39 M. R. Banaji, 'Implicit attitudes can be measured', in H. L. Roediger III, J. S. Nairne, I. Neath, & A. Surprenant (eds.), *The nature of remembering: Essays in honor of Robert G. Crowder* (Washington, DC: American Psychological Association, 2001), pp. 117–150. Quoted in (Banaji, Nosek, & Greenwald, 2004), p. 284.

40 (Silverberg, 2006), p. 3.

41 (Grossi, 2008), p. 100.

42 (Fausto-Sterling, 2000), p. 118.

43 (Rivers & Barnett, 2007), para. 4.

44 See (Fine, 2008).

45 Quoted in (Pierce, 2009), para. 8.

46 A point made, for example, by (Bleier, 1984). She suggests that 'Paradoxically, it is not our brains or our biology but rather the cultures that our brains have produced that constrain the nearly limitless potentialities for behavioral flexibility provided us by our brains.' (p. viii).

1. WE THINK, THEREFORE YOU ARE

1 (Morris, 1987), p. 140.

2 Sociologists Cecilia Ridgeway and Shelley Correll point out that there is

something curious about how our gender beliefs can be so narrow 'since no one ever has the experience of interacting with a concrete person who is just a man or just a woman in a way that is not affected by a host of other attributes such as the person's race or level of education.' (Ridgeway & Correll, 2004), p. 513.

3 See (Rudman & Glick, 2008), chapter 4. This book provides a compelling and comprehensive account of the social psychology of gender.

4 (Ridgeway & Correll, 2004), p. 513. Much of the research discussed in this book, it should be acknowledged, is restricted to the white, middle-class, heterosexual wedge of society. But then, it is the disparity between the male and female halves of this privileged group that is most likely to be taken as evidence for the 'naturalness' of gender roles.

5 For overview see (Nosek, 2007a).

6 (Nosek & Hansen, 2008), p. 554, references removed.

7 For theoretical discussions, see for example (Gawronski & Bodenhausen, 2006; Smith & DeCoster, 2000; Strack & Deutsch, 2004).

8 For example (Banaji & Hardin, 1996). For brief overview see (Bargh & Williams, 2006).

9 To experience the Implicit Association Test yourself, and find out more about it, visit Harvard University's Project Implicit Web site: http://implicit.harvard.edu/implicit/.

10 (Rudman & Kilianski, 2000).

11 Brian Nosek notes that correlations between implicitly measured social attitudes (such as towards minority groups) and self-reported attitudes are especially weak when participants are highly egalitarian university students, whereas in less egalitarian groups the relationships are stronger (Nosek, 2007a). The nature of the relationship between explicit and implicit attitudes and other constructs – to what extent are they distinct? – is still not clear, and subject to debate.

12 For example (Mast, 2004; Nosek et al., 2009; Rudman & Kilianski, 2000).

13 (Dasgupta & Asgari, 2004).

14 For example (Kunda & Spencer, 2003). Or see (Fine, 2006).

15 I was alerted to this quotation, in the context of understanding the self, in an interview with Brian Nosek.

16 See (Wheeler, DeMarree, & Petty, 2007).

17 This is especially predicted by John Turner's self-categorisation theory, which is most explicit in distinguishing between personal identity and social identity. While both self-categorisation theory and the active-self account (and other similar models, such as the notion of a working self-concept) regard the self as dynamic and context-dependent, self-categorisation theory proposes that 'the self should not be equated with enduring personality structure' because an

infinite number of different social identities could become active, depending on the social context (Onorato & Turner, 2004), p. 259. Evidence for self-stereotyping under conditions of gender salience comes, for example, from (Hogg & Turner, 1987; James, 1993).

18 (Chatard, Guimond, & Selimbegovic, 2007).

19 Quoted in (Horne, 2007).

20 (Sinclair, Hardin, & Lowery, 2006).

21 (Steele & Ambady, 2006).

22 (Steele & Ambady, 2006), p. 434.

23 (Garner, 2004), p. 177.

24 William James (1890), *The Principles of Psychology*, p. 294. Quoted in p. 529 of (Sinclair, Hardin, & Lowery, 2006).

25 (Sinclair, Hardin, & Lowery, 2006; Sinclair et al., 2005; Sinclair & Lun, 2006), p. 529.

26 (Davies, 1989), p. 17.

27 (Galinsky, Wang, & Ku, 2008).

28 (Sinclair et al., 2005).

29 For a sociological perspective on this idea, see (Paechter, 2007).

2. WHY YOU SHOULD COVER YOUR HEAD WITH A PAPER BAG IF YOU HAVE A SECRET YOU DON'T WANT YOUR WIFE TO FIND OUT

1 (Brizendine, 2007), p. 161.

2 A claim made in the blurb of Brizendine's book.

3 (Baron-Cohen, 2003), p. 2.

4 The Autism Research Centre was the source of the Empathy Quotient and Systemizing Quotient questionnaires: http://www.autismresearchcentre.com/tests/default.asp.

5 See (Baron-Cohen, Knickmeyer, & Belmonte, 2005), table 1, p. 821. Sixty percent of men report an S-type brain, compared with 17 percent of women. (Percentages include 'extreme' E and S brain types.)

6 (Schaffer, 2008), entry 3 ('Empathy queens'), para. 5.

7 (Eisenberg & Lennon, 1983).

8 Quoted in (Schaffer, 2008), entry 3 ('Empathy queens'), para. 8.

9 (Davis & Kraus, 1997), p. 162.

10 (Ames & Kammrath, 2004), p. 205; (Realo et al., 2003), p. 434.

11 (Voracek & Dressler, 2006).

12 Both the EQ and the Reading the Mind in the Eyes test, also from Simon Baron-Cohen's lab, ask participants to state their sex before beginning the

questionnaire. As will become clear later in the chapter, it's possible that the correlation between the two arises because the salience of gender-related norms increases both self-reported empathy and empathic performance, to a greater or lesser degree in different participants.

13 (Ickes, 2003), p. 172.

14 (Levy, 2004), p. 322.

15 (Voracek & Dressler, 2006). If you used information about whether someone scored below or above average on the test to try to guess his or her sex you would be correct barely more often than chance.

16 These and further details of the PONS and its interpretation, as well as the IPT, are summarised in (Graham & Ickes, 1997). To give you an idea of the size of the gender difference on the PONS, which Graham and Ickes describe as 'respectable' (p. 123), the average woman on this test (scoring at the 50th percentile) is equivalent to a slightly superior man (scoring at the 66th percentile for the male population). In their discussion of the greater female advantage for 'leaky' channels of communication, they are referring to the work (and term) of R. Rosenthal and B. DePaulo, 'Sex differences in eavesdropping on nonverbal cues', *Journal of Personality and Social Psychology* 37 (1979), pp. 273–285.

17 (Brizendine, 2007), p. 160.

18 This hypothesis again refers to the work of Rosenthal & DePaulo, cited in (Graham & Ickes, 1997).

19 (Graham & Ickes, 1997), p. 126.

20 (Ickes, 2003), quotations from pp. 125 and 126, respectively.

21 (Ickes, Gesn, & Graham, 2000).

22 (Ickes, 2003), p. 135.

23 (Klein & Hodges, 2001). Men also scored equivalently to women when the sympathy rating was requested *after* the empathic accuracy test.

24 (Thomas & Maio, 2008), p. 1173. This effect was only found for an easy-to-read target, not a difficult-to-read target.

25 (Koenig & Eagly, 2005), p. 492.

26 (Marx & Stapel, 2006c), p. 773.

27 (Seger, Smith, & Mackie, 2009), p. 461.

28 (Ryan, David, & Reynolds, 2004). Gilligan's work and critiques summarised here also.

29 This claim also found support in (Ryan, David, & Reynolds, 2004), study 1.

30 (Ryan, David, & Reynolds, 2004), pp. 253 and 254, respectively, references removed.

3. 'BACKWARDS AND IN HIGH HEELS'

1 For meta-analysis, see (Voyer, Voyer, & Bryden, 1995).

2 (Moore & Johnson, 2008; Quinn & Liben, 2008). It's worth noting that the early appearance of this difference does not necessarily mean that experiential factors could not be responsible. For example, male babies could be given more gross stimulation that stimulates visuospatial skills. Interestingly, one study found that boys and girls from a low socioeconomic background underperformed equally on a visuospatial task, whereas more-privileged boys outperformed their female counterparts. This points towards the importance of experiential factors in male advantage (Levine et al., 2005). Moreover, an early advantage for males doesn't mean that this must inevitably persist. In other cognitive domains, gender differences are transient.

3 Needless to say, this is a complex issue. As Nora Newcombe recently summarised it, not only do men, on average, outperform women on mental rotation tasks, particularly at the highest levels, spatial visualisation skills are relevant to success in fields such as physics, mathematics, computer science and engineering. However, as she also notes, there are difficulties with the argument that these genuine sex differences are biologically caused and immutable. With regard to the first point – biological causation – she notes that hypotheses attempting to account for biological mechanisms have not been successful. (Two of these, hormonal accounts and sex differences in lateralisation, are discussed in the second part of the book. The other ideas – an X-linked recessive gene for spatial ability and males' later puberty – have not been supported by the evidence.) Newcombe also notes that, despite superficial plausibility, evolutionary explanations entail numerous untested assumptions. One further important point raised by Newcombe is whether extra increments in mental rotation ability are important, beyond some high threshold. (As Amanda Schaffer dryly put it in *Slate*, 'when it comes to the diverse precincts of high-level science, spatial reasoning only gets you so far. Rock-star academics don't necessarily spend their days turning geometric figures around in their minds.') Newcombe points out that '[t]hinking creatively, explaining one's data, or inspiring a research team may be pretty important as well!' (Newcombe, 2007), p. 75. A recent, very comprehensive review of 'sociocultural and biological considerations' with respect to women's underrepresentation in science concluded that the 'process needed to establish male advantage in STEM fields as a function of superior spatial ability (possibly because of its role in advanced mathematics) is littered with loopholes. Nothing close to a tightly reasoned and supported argument currently exists.' (Ceci, Williams, & Barnett, 2009), p. 250.

4 Reviewed, for example, by (Newcombe, 2007). Recent studies have also found that playing computer games improves mental rotation ability, and in women more so than in men (Cherney, 2008; Feng, Spence, & Pratt, 2007).

5 (Sharps, Price, & Williams, 1994). Task instructions quoted from pp. 424 and 425. Men in the masculine condition outperformed men and women in all other groups. See also (Moè & Pazzaglia, 2006).

6 (McGlone & Aronson, 2006).

7 (Hausmann et al., 2009).

8 (Moè, 2009).

9 M. B. Ritter, *More than gold in California 1849–1933* (Berkeley, CA: The Professional Press, 1933), p. 161. Quoted in (Morantz-Sanchez, 1985), p. 118.

10 C. M. Steele, S. J. Spencer, & J. Aronson, 'Contending with group image: The psychology of stereotype and social identity threat'. In M. P. Zanna (ed.), *Advances in experimental social psychology*, vol. 34 (San Diego: Elsevier, 2002), p. 385. Quoted in (Shapiro & Neuberg, 2007), p. 109.

11 Readers interested in reading more about stereotype threat are strongly recommended to visit the Web site http://reducingstereotypethreat.org, authored by social psychologists Steven Stroessner and Catherine Good, which provides detailed and comprehensive coverage of the academic literature.

12 (Good, Aronson, & Harder, 2008).

13 For example (Marx & Stapel, 2006b; Marx, Stapel, & Muller, 2005; Thoman et al., 2008).

14 (Good, Aronson, & Harder, 2008), p. 25.

15 (Walton & Spencer, 2009), p. 1133. Although they note that their samples may not be representative of the general population, their effect sizes suggest that the SAT Maths may underestimate women's abilities by about 20 points (compared with a gender gap of 34 points). For African and Hispanic Americans, SAT Reading tests may underestimate ability by about 40 points.

16 For example (Adams et al., 2006; Danaher & Crandall, 2008; Davies et al., 2002; Inzlicht & Ben-Zeev, 2000; Logel et al., 2009).

17 See (Nguyen & Ryan, 2008).

18 (Marx, Stapel, & Muller, 2005).

19 For example (Cadinu et al., 2003; Stangor, Carr, & Kiang, 1998) and (Marx & Stapel, 2006a), p. 244. As David Marx has argued, and his work has been demonstrating, priming a self-relevant stereotype has effects different from, and greater than, standard stereotype priming effects.

20 (Cadinu et al., 2005), p. 574.

21 (Logel et al., 2008). See also (Davies et al., 2002) who found that gender stereotypes were activated in women who saw gender-stereotyped

advertisements, compared with controls, and that this activation predicted maths underperformance.

22 (Beilock, Rydell, & McConnell, 2007; Schmader & Johns, 2003). For review see (Schmader, Johns, & Forbes, 2008).

23 (Johns, Inzlicht, & Schmader, 2008).

24 For example (Aronson et al., 1999; Croizet et al., 2004).

25 Presenting the test as gender-neutral (i.e., males and females score equally) enhances women's performance (for example [Spencer, Steele, & Quinn, 1999]), and does not have the same detrimental effect on working memory (for example [Johns, Inzlicht, & Schmader, 2008]).

26 (Seibt & Förster, 2004).

27 (Gladwell, 2008), pp. 87, 87, and 88, respectively.

28 See (Nguyen & Ryan, 2008) who concluded from their meta-analysis that low maths-identified women are the least affected by stereotype threat. Interestingly, they found that moderately identified women are the most affected (more so than high-identified females), although they note that there is some inconsistency in how 'identification' is defined and operationalised.

29 For instance (Beilock, Rydell, & McConnell, 2007) found that stereotype threat most affects maths problems that rely more heavily on working-memory resources.

30 These numbers, from the National Science Foundation, are cited in (Ceci, Williams, & Barnett, 2009).

31 (Inzlicht & Ben-Zeev, 2000).

32 (Schmader, Johns, & Barquissau, 2004).

33 (Kiefer & Sekaquaptewa, 2007).

34 See (Blanton, Crocker, & Miller, 2000; Marx, Stapel, & Muller, 2005). For effect of 'closeness' of the model, see (Marx et al., unpublished manuscript), who found that women exposed to a highly maths-competent, socially 'close', female role model performed better on a maths test than women exposed to a socially 'distant', but equally competent, female role model. (Lockwood, 2006) found that women in particular benefit by having an inspiring female role model. In general, research into social comparison processes finds that, among other factors, our self-evaluations and behaviour are more likely to assimilate to another person to the extent that we feel psychologically similar to them. Otherwise the standard set by the other person becomes a contrast against which our own self-evaluation and behaviour reacts. See, for example (Mussweiler, Rüter, & Epstude, 2004).

35 (Marx & Roman, 2002; McIntyre et al., 2005; McIntyre, Paulson, & Lord, 2003).

36 (Josephs et al., 2003; Newman, Sellers, & Josephs, 2005).

37 See (Rogers, 1999), pp. 75–85. It's also worth noting that although some have

argued that the relationship between testosterone levels and competition is different in women and men, there are currently too few studies available with women to draw a fair comparison. See (van Anders & Watson, 2006), pp. 215–220.

38 (Sherwin, 1988).

39 (Rogers, 1999), p. 83.

40 (Josephs et al., 2003), p. 162.

41 (Huguet & Régner, 2007; Neuville & Croizet, 2007). Also (Ambady et al., 2001) found stereotype threat effects in lower-elementary and middle school girls, although unexpectedly upper-elementary girls did better when gender identity was salient.

42 (Nosek et al., 2009), p. 10597. These relationships held even controlling for a general indicator of social gender inequality.

4. I DON'T BELONG HERE

1 (Hines, 2004), p. vii.

2 (Haslanger, 2008), p. 211.

3 Quoted in (McCrum, 2008) p. 22.

4 (Mullarkey, 2004), pp. 369 and 370, respectively.

5 (Mullarkey, 2004), pp. 373 and 374.

6 (Pinker, 2008), p. 5.

7 (Steele, 1997), p. 618.

8 (Murphy, Steele, & Gross, 2007).

9 (Davies et al., 2002).

10 (Davies, Spencer, & Steele, 2005).

11 (Gupta & Bhawe, 2007), p. 74.

12 A point made by (Cheryan et al., 2009).

13 (Cheryan et al., 2009).

14 I. J. Seligsohn, *Your Career in Computer Programming* (New York: Simon & Schuster, 1967), cited in (Gürer, 2002a), p. 176.

15 (Gürer, 2002b), p. 120.

16 Sapna Cheryan, personal communication, November 25, 2009.

17 (Cheryan et al., 2009), p. 1058.

18 (Spelke & Grace, 2006), p. 726.

19 The criteria were changed to downplay prior programming ability – which was shown not to be a predictor of success in the CS major, and instead focus on 'indicators of future visionaries and leaders in computer science.' (Blum & Frieze, 2005), p. 117. The study referred to was conducted by Jane Margolis

and Allen Fisher, reported in *Unlocking the Clubhouse* (Cambridge, MA: MIT Press, 2002).

20 (Blum & Frieze, 2005), quotations from pp. 113 and 114.

21 (Good, Rattan, & Dweck, unpublished).

22 (Haslanger, 2008), p. 212.

23 (Correll, 2001), p. 1724.

24 (Correll, 2004), p. 102.

25 (Pronin, Steele, & Ross, 2004).

26 The article was C. P. Benbow and J. C. Stanley, 'Sex differences in mathematical ability: fact or artifact?' *Science* 210 (1980), pp. 262–1264.

27 Quoted in (Pronin, Steele, & Ross, 2004), p. 159.

28 (Hewlett, Servon et al., 2008), p. 11.

29 (Hewlett, Luce, & Servon, 2008), p. 114.

30 (Hewlett, Servon et al., 2008), quotations from pp. 11 and 12.

31 For instance, from their comprehensive review of possible biological and social factors contributing to female underrepresentation in science, Stephen Ceci and colleagues conclude that the evidence for the role of biological factors is 'contradictory and inconclusive.' They suggest that the evidence points most strongly to the role of women's preferences – which they note could either be seen as free choices or constrained 'choices' – with a secondary factor being poorer performance on gatekeeper tests, which they regard as being more likely due to sociocultural than biological factors (Ceci, Williams, & Barnett, 2009), p. 218.

5. THE GLASS WORKPLACE

1 (Fara, 2005). See pp. 93–96.

2 (Barres, 2006), p. 134.

3 (Schilt, 2006), p. 476.

4 See also data and arguments provided by (Valian, 1998).

5 (Steinpreis, Anders, & Ritzke, 1999). Estimated from figure 5, p. 520.

6 (Steinpreis, Anders, & Ritzke, 1999), p. 523.

7 (Davison & Burke, 2000). However, sex discrimination was greater when less job-relevant information was available.

8 (Heilman, 2001), p. 659.

9 (Biernat & Kobrynowicz, 1997).

10 Interestingly, when evaluations were made on vague, subjective scales (*very poorly to very well*, or *very unlikely to very likely*), Katherine was preferred for the chief of staff position, while Kenneth was favoured as a secretary. However, the researchers suggested that this was because Katherine was seen as being a good

candidate for the masculine job *for a woman*, while Kenneth was perceived as an impressive potential secretary *for a man*. When more objective scales were used that forced the raters to put numbers and percentiles to their evaluations, the pattern reversed.

11 (Correll, Benard, & Paik, 2007). Participants were undergraduates, told that their input would be used along with other information in real hiring decisions.

12 See, for example (Crosby, Williams, & Biernat, 2004) and other articles in the same issue.

13 (Bledsoe, 1856), pp. 224 and 225.

14 (Rudman & Kilianski, 2000).

15 See, for example (Rudman, 1998; Rudman & Glick, 1999, 2001). For summary of research suggesting that warmth and competence are fundamental dimensions of social perception, see (Fiske, Cuddy, & Glick, 2007).

16 A phrase coined by Janet Holmes, author of *Gendered Talk*, cited by (Cameron, 2007), p. 141.

17 M. Dowd, 'Who's hormonal? Hillary or Dick?' *New York Times*, February 8, 2006, p. A21, quoted by study authors (Brescoll & Uhlmann, 2008), p. 268.

18 (Cuddy, Fiske, & Glick, 2004). Interestingly, in this study gender per se was not a factor for discrimination, although it's possible that this was because of the phenomenon described in note 10.

19 (Rudman et al., manuscript submitted for publication).

20 (Norton, Vandello, & Darley, 2004).

21 (Uhlmann & Cohen, 2005), p. 479.

22 (Uhlmann & Cohen, 2005), p. 478, references removed.

23 (Phelan, Moss-Racusin, & Rudman, 2008), p. 408.

24 Quoted in (Monastersky, 2005), para. 42.

25 (Bolino & Turnley, 2003; Bowles, Babcock, & Lai, 2007; Butler & Geis, 1990; Heilman & Chen, 2005; Heilman et al., 2004; Sinclair & Kunda, 2000).

26 (Heilman, 2001), p. 670.

27 (Cameron, 2007), pp. 134 and 135.

28 (Ryan et al., 2007), p. 270.

29 See (Ashby, Ryan, & Haslam, 2007; Haslam & Ryan, 2008). Other data forthcoming, summarised in (Ryan et al., 2007), pp. 270 and 271.

30 (Uhlmann & Cohen, 2005).

31 (Williams, 1992), p. 256.

32 (Wingfield, 2009).

33 (Gorman & Kmec, 2007), p. 839.

34 Quoted in (Allen, 2009), para. 7.

35 (Hersch, 2006), p. 352.

36 (Liben, Bigler, & Krogh, 2001).

6. XX-CLUSION AND XXX-CLUSION

1 Quoted in (MacAdam, 1914), para. 12.

2 (Glick & Fiske, 2007), p. 162.

3 (Glick & Fiske, 2007), p. 163.

4 Quoted in (MacAdam, 1914), para. 13.

5 (Selmi, 2005), pp. 41 and 25, respectively.

6 (Selmi, 2005), p. 31.

7 (Roth, 2004), p. 630.

8 (Hewlett, Servon et al., 2008), pp. 7 and 8, respectively.

9 Quoted in (Verghis, 2009), p. 26.

10 (Morgan & Martin, 2006), p. 121.

11 (Morgan & Martin, 2006), quotations from pp. 116, 117, and 118, respectively.

12 Quoted in (Dugan, 2008).

13 Quoted in D. Valler, Business visitors expect this on the agenda. *Coventry Evening Telegraph*, November 9, 2005, p. 8. Quoted in (Jeffreys, 2008), p. 166.

14 Quoted in (Lynn, 2006), para. 22.

15 http://www.stringfellows.co.uk/clubs/pages/corporate-events.php, accessed on August 27, 2009.

16 (Barnyard & Lewis, 2009).

17 (Morgan & Martin, 2006), p. 117.

18 According to the court testimony of one London financial executive, cited by (Lynn, 2006).

19 (Jeffreys, 2008), p. 155.

20 (Lynn, 2006), para. 24.

21 (Selmi, 2005), pp. 24 and 36, respectively. Selmi makes this argument in the context of a discussion of changing academic theories of sexual harassment.

22 (Selmi, 2005), p. 7.

23 (Hewlett et al., 2008), p. 7.

24 (Hinze, 2004), p. 105, referring to the Equal Employment Opportunity Commission definition of a hostile environment.

25 (Hinze, 2004), pp. 109, 111, and 111, respectively.

26 (Hinze, 2004), pp. 120, 114, 115, 114–115, and 115, respectively.

27 (Kimmel, 2008), p. 227.

28 (Woodzicka & LaFrance, 2001). Sixty-eight percent of women who merely *imagined* themselves in this situation thought that they would refuse to answer at least one question, 16 percent said they would leave the interview, and 6 percent said they would report the interviewer to his supervisor. The percentages of women who actually responded in these ways to *real* sexually

harassing interview questions were, respectively, 0 percent, 0 percent and 0 percent.

29 *Philadelphia Evening Bulletin*, November 8, 1869. Quoted in (Morantz-Sanchez, 1985), p. 9.

30 (Selmi, 2005), p. 25 then p. 30.

31 (Gutek & Done, 2001).

7. GENDER EQUALITY BEGINS (OR ENDS) AT HOME

1 M. Ulrich, 'Men are queer that way: Extracts from the diary of an apostate woman physician', *Scribner's Magazine* 93 (June 1933), pp. 365–369. Quoted in (Morantz-Sanchez, 1985), pp. 325 and 326 (epigraph included).

2 (Hochschild, 1990).

3 See, for example (Bittman et al., 2003; Brines, 1994).

4 Quoted in (Belkin, 2008), para. 28.

5 (Gray, 2008), quotations from pp. 123, 123, 124, 125, 123, and 123, respectively.

6 (Gurian, 2004), pp. 219, 219, and 220, respectively.

7 (Bittman et al., 2003), p. 198. Note that sociologists are not entirely in agreement as to how this pattern is best explained.

8 The phrase 'doing gender' refers to a theory by sociologists Candace West and Don Zimmerman.

9 (Tichenor, 2005), pp. 197, 198, 199, 201, and 199–200, respectively.

10 (Selmi, 2008), p. 21.

11 Ulrich, 'Men are queer that way'. Quoted in (Morantz-Sanchez, 1985), p. 326.

12 (Hochschild, 1990) also discusses the contradictions between people's explicit and implicit gender ideologies.

13 (Devos et al., 2007).

14 See (Greenwald et al., 2009).

15 (Rudman & Heppen, 2003).

16 (Rudman, Phelan, & Heppen, 2007).

17 (Stone, 2007), p. 64.

18 For interesting discussions of this issue, see (Jolls, 2002; Selmi, 2008).

19 Quoted in (Belkin, 2008), para. 39.

20 For example, see (Jolls, 2002) for evidence of discrimination with implications for wages, (Weichselbaumer & Winter-Ebmer, 2005) for data on the international gender wage gap, and (Kilbourne et al., 1994) for data showing

that occupations pay less to the extent that they have a higher proportion of female workers or involve greater nurturing.

21 In fact, sociological studies of how gender ideology changes in response to life experience find that parenthood doesn't inevitably bring about less egalitarian views. People who have children at a nonnormative time don't show this shift, and parenthood brings about a shift towards more egalitarian views in unmarried parents (Davis, 2007; Vespa, 2009).

22 Ulrich, 'Men are queer that way'. Quoted in (Morantz-Sanchez, 1985), p. 327.

23 (Brizendine, 2007), pp. 151 and 208, respectively.

24 (Brizendine, 2007), p. 207.

25 (Stone, 2007), pp. 77 and 78.

26 See (van Anders & Watson, 2006). Also (Silvers & Haidt, 2008) who found that watching a morally elevating video triggered nursing in mothers, suggesting oxytocin release.

27 See (van Anders & Watson, 2006; Wynne-Edwards, 2001; Wynne-Edwards & Reburn, 2000).

28 (Deutsch, 1999), p. 230. First quotation is Deutsch, second quotation is from her interviewee.

29 (Rosenblatt, 1967).

30 Wynne-Edwards suggests that 'paternal and maternal behavior are homologous at a neural and an endocrine level', and that this makes sense for reasons of parsimony (Wynne-Edwards, 2001), p. 139.

31 (Demos, 1982), p. 429. See also (Collins, 1982).

32 *Parents' Magazine*. Family prayer in men of business. May 1842, p. 198. Quoted in (Demos, 1982), p. 436.

33 See, for example, discussion in (Hamilton, 2004), pp. 205–207. The *Yearning for Balance* report cited by Hamilton found that 40 percent of 'downshifters' (that is, people who shift their emphasis to leisure and relationships rather than economic success) in a survey of 800 adults were men. The Harwood Group, *Yearning for Balance: Views of Americans on consumption, materialism, and the environment*, prepared for the Merck Family Fund. http://www.iisd. ca/consume/harwood.html, accessed on August 27, 2009.

34 Quoted in (Montemurri, 2009), para. 3.

8. GENDER EQUALITY 2.0?

1 (Pinker, 2008), p. 255.

2 (Levy, 2004), p. 323.

3 The cartoon is by Tom Cheney, published in *The New Yorker* on May 3, 1993.

4 (Hamilton, 2004), p. 130. Hamilton is not referring here to gender, but to the

role of marketing and a political emphasis on the primacy of the importance of economic growth on people's preferences.

5 (Mason & Goulden, 2004).

6 (Gharibyan, 2009).

7 (Gharibyan, 2007, p.10; Gharibyan & Gunsaulus, 2006). Computer science is not male dominated in Singapore or Malaysia either (Galpin, 2002).

8 (Charles & Bradley, 2009).

9 (Peplau & Fingerhut, 2004).

10 (Charles & Bradley, 2009), p. 929.

11 For example (Costa, Terracciano, & McCrae, 2001; Fullagar et al., 2003; Guimond, 2008; Prime et al., 2008).

12 (Steele & Ambady, 2006), pp. 434 and 435.

13 (Ridgeway & Correll, 2004), p. 520.

9. THE 'FETAL FORK'

1 (Hess, 1990), p. 81, references removed.

2 (Brizendine, 2007), pp. 36, 36, and 37 and 38, respectively.

3 I say 'seems' because, so far as I can tell, Brizendine does not refer to any evidence that supports these terrifying claims. In the notes, to support the claim about 'growing more cells in the sex and aggression centres', Brizendine cites an irrelevant review of cortical development in the rat (M. Sur and J. L. Rubenstein, 'Patterning and plasticity of the cerebral cortex', *Science* 310, no. 5749 [2005], pp. 805–810), which makes no mention of sex differences. To support the claim that '[t]he fetal girl's brain cells sprout more connections in the communication centers and areas that process emotion' she refers the reader to Chapter 6, 'Emotions'. However, I was unable to find any research or discussion of foetal brain development in this chapter. The absence of support for these and other similar claims is discussed by Mark Liberman. See http://158.130.17.5/~myl/languagelog/archives/003541. html and http://158.130.17.5/~myl/languagelog/archives/004694.html, both accessed on October 5, 2009.

4 As noted by Mark Liberman, http://itre.cis.upenn.edu/~myl/languagelog/archives/003551.html, accessed September 16, 2009.

5 (Baron-Cohen, 2009), para. 5 and para. 22 respectively.

6 This section summarised from (Hines, 2004).

7 The mechanism and threshold of necessary testosterone, and the timing of the critical period, are different for the internal reproductive organs and the external genitalia.

8 For overview see (Morris, Jordan, & Breedlove, 2004).

9 A useful summary is provided by (Breedlove, Cooke, & Jordan, 1999).

10 In one species of bird, the African bush shrike, males have superior vocal control areas (that is, 'larger nuclei, denser connections, more synapses, *etc.*') even though the complexity of male and female songs is identical. Implication? 'The link between song production and size of the vocal control nuclei may not be as simple as it first appeared.' (De Vries, 2004), p. 1063.

11 See, for example, Mark Liberman's discussion of Leonard Sax's use of data on rat vision to draw conclusions about human gender difference and single-sex schooling (http://itre.cis.upenn.edu/~myl/languagelog/archives/003473.html).

12 (Hines, 2004), p. 82.

13 These include different timing, different physiological effects and different hormonal mechanisms. For example, while injecting testosterone into female rats soon after birth disrupts the estrous cycle (the rat version of the menstrual cycle), prenatal testosterone during the equivalent critical period in humans and other primates doesn't have the same disrupting effect. Also, the role of testicular hormones converted to oestrogens in sexual differentiation may be different in rats and primates.

14 For example, Wallen argues that 'the dominant rat and mouse models of sexual differentiation seem unlikely to apply to human sexual differentiation.' (Wallen, 2005), p. 8.

15 See (Wallen, 1996). Referring to frequency of 'threat' behaviour such as baring teeth and staring.

16 For high doses of prenatal testosterone treatment, late in gestation. Earlier in gestation, no effect of the same high dose of testosterone on rough play is seen. In both rats and rhesus monkeys, prenatal androgen treatment also affects sexual behaviour, for example, degree of mounting. See (Wallen, 2005).

17 Early blocking using flutamide reduces the masculinisation of the genitalia and results in rough play and mounting intermediate between male and female behaviour. Late blocking reduces penis length, has no effect on rough play (even though in females it is testosterone *late* in gestation that appears to be important in influencing rough play) and actually *increases* mounting behaviour, which is opposite of what one would expect (Wallen, 2005).

18 Described in (De Vries, 2004).

19 Quoted in (Kolata, 1995), para. 22. Gorski adds that 'nothing like it has been shown in humans.'

20 (De Vries, 2004), p. 1064.

21 For example, a book for parents published by the Gurian Institute claims that '[w]ithout the testosterone hits received *in utero* by her male counterparts, her brain continued on the female default path, providing specialized circuitry for

communication, emotional memory, and social connection.' (Gurian Institute, Bering, & Goldberg, 2009), p. 32.

22 For valuable discussions of the problems with the orthodox view of the organisational/activational hypothesis, see (Breedlove, Cooke, & Jordan, 1999; Fausto-Sterling, 2000; Kaplan & Rogers, 2003; Moore, 2002; Rogers, 1999).

23 (Moore, 2002), pp. 65 and 66.

24 (Moore, Dou, & Juraska, 1992).

25 (Moore, 2002), p. 65.

26 (Barnett & Rivers, 2004), p. 200. For criticism of the lack of impact of this important work in the scientific community, see (Kaplan & Rogers, 2003), pp. 53–56.

27 (Geschwind & Behan, 1982).

28 Quoted in (Kolata, 1983), p. 1312.

29 Note, according to the model, extremely high levels of foetal testosterone will have detrimental effects on right-hemisphere development, and thus visuospatial function. As several researchers and commentators have pointed out, there is little in the way of evidence for the model, yet despite this it enjoys tremendous scientific and popular appeal and influence. In particular, Ruth Bleier has made an excellent critique of the model, and her criticisms and data have also been well summarised by Carol Tavris (Bleier, 1986; Tavris, 1992). For further critiques see also (Fausto-Sterling, 1985; Grossi, 2008; Nash & Grossi, 2007; Rogers, 1999). A comprehensive account of the data with regard to the Geschwind-Behan-Galaburda model, as it is more formally known, which proposes a link among foetal testosterone, left-handedness, giftedness and immune-system functioning, concluded that '[a]n overall evaluation of the model suggests that it is not well supported by empirical evidence and that in the case of several key theoretical areas, the evidence that does exist is inconsistent with the theory.' (Bryden, McManus, & Bulman-Fleming, 1994), p. 103.

30 (Bleier, 1986).

31 (Gilmore et al., 2007), who found that, contrary to adults and older children, in neonates of both sexes the left hemisphere is larger than the right. See also (Nash & Grossi, 2007), p. 15, for discussion of lack of support for the model in studies of adult brains. This is in contrast to research with rats, which has demonstrated the relatively larger right hemisphere in males and the dependence of this on neonatal testosterone (Diamond, 1991). Note that Diamond's summary of this work also points to the importance of experiential factors in hemisphere asymmetry. I do not know whether researchers have investigated whether the effect of neonatal testosterone on cerebral lateralisation occurs directly and/or via the different social experiences triggered by higher neonatal

testosterone – a possibility suggested by the work of Celia Moore described earlier.

32 As Baron-Cohen puts it, 'the more you have of this special substance [testosterone, especially early in development], the more your brain is tuned into systems and the less your brain is tuned into emotional relationships.' (Baron-Cohen, 2003), p. 105. It's not clear that 'extreme male' is a good description of the profile of people with autism. You'll remember from the first part of the book that empathy can be either cognitive (mind reading) or affective (sympathy). In seminal work, Simon Baron-Cohen showed that people with autism struggle with cognitive empathy, that is, they can't seem to read other people's intentions, beliefs and feelings with the intuitive ease that most of us enjoy (Baron-Cohen, 1997). Yet several strands of research now suggest that people with autism don't lack *affective* empathy (Blair, 1996; Dziobek et al., 2008; Rogers et al., 2007). This is problematic for Baron-Cohen's thesis because, as Levy has pointed out in (Levy, 2004), according to Baron-Cohen (see [Baron-Cohen, 2003], p. 120) the typical male profile is the precise opposite. Baron-Cohen suggests that men's empathy disadvantage is greater for affective, rather than cognitive, empathy, the latter being vital for success in domains of predominantly male achievement. (Think how badly a poor mind reader would get on in business, politics or law.) It's also worth noting the possibility that high foetal testosterone 'reduces the threshold at which autistic symptoms manifest', rather than causing autistic symptoms directly, as suggested by (Skuse, 2009), p. 33.

10. IN 'THE DARKNESS OF THE WOMB'
(AND THE FIRST FEW HOURS IN THE LIGHT)

1 (Gurian Institute, Bering, & Goldberg, 2009), pp. 18 and 19.

2 Remember Celia Moore's work, which found that early testosterone affected the mother rat's behaviour. Foetal testosterone levels might affect, say, the physical appearance of the child in some way that influences how the child is treated (for example, by making the face more masculine). It's also possible that parents who have children with higher levels of foetal testosterone tend to be different from those who don't, in some way that affects the environment they provide to their children.

3 With respect to the use of maternal testosterone (mT), one clinical study that measured foetal testosterone directly did find that it correlated with mT (Gitau et al., 2005). However, as noted by van de Beek et al. mT levels are not higher in women carrying boys than in those carrying girls, which suggests 'that maternal

serum androgen levels are not a clear reflection of the actual exposure of the fetus to these hormones.' (van de Beek, et al., 2004), p. 664. Also, testosterone can only act on the brain if it is free (that is, if the testosterone is not bound to another molecule). One way this can be indirectly assessed is to also measure levels of SHBG (sex hormone binding globulin). The more SHBG, the less free testosterone is likely to be available. The two studies that used maternal serum measured both. One found a correlation between a sex-typed behaviour measure and mT but not SHBG (Hines et al., 2002). The other found a correlation with SHBG but not mT (Udry, 2000). There therefore seems some uncertainty as to which (if either) is the appropriate proxy for foetal testosterone (fT) exposure. For amniotic testosterone (aT) 'there is no direct evidence to either support or contradict' the assumption that aT is correlated with the levels of testosterone acting on the foetal brain (Knickmeyer, Wheelwright et al., 2005), p. 521. (van de Beek et al., 2004) suggest aT as the best index of fT exposure, but they also acknowledge the lack of much understanding of the relationship between levels of testosterone in the amniotic fluid – the main source of which is foetal urine – and in the foetal blood. Van de Beek and colleagues note that 'there is no hard evidence of a direct relationship between amniotic testosterone and fetal serum testosterone.' (van de Beek et al., 2009), p. 8. Finally, the use of the digit ratio as a marker of prenatal testosterone exposure is controversial and lacks clear empirical support. For review see (McIntyre, 2006). One researcher has complained that '[t]he lightheartedness of using certain biological markers in adulthood as indicators of prenatal androgen exposure is not warranted.' (Gooren, 2006), p. 599. Because digit ratio seems to be the most controversial index of prenatal androgen exposure within this field of interest, I don't attempt here to provide anything like a comprehensive account of research findings using this technique.

4 I am very grateful to Giordana Grossi for her helpful discussions of the following literature.

5 It's important that correlations are seen within sex. Otherwise gender socialisation might create psychological differences that then correlate with foetal testosterone for the simple reason that boys have higher foetal testosterone than girls.

6 (Lutchmaya, Baron-Cohen, & Ragatt, 2002). The data from this study are not completely straightforward. For boys and girls together, amniotic testosterone (aT) did indeed correlate negatively and linearly with frequency of eye contact. That is, children with high aT had lower eye contact frequency than children with low aT. However, there was also a quadratic relationship meaning that eye contact frequency decreased with increasing aT in the low aT range (as predicted), but *increased* with increasing aT in the high aT range. This same

pattern appeared when looking at boys separately. In girls only, no relationship at all was seen between aT and eye contact frequency. These data, then, are not consistent with the claim that 'the higher your levels of pre-natal testosterone, the less eye contact you now make' (Baron-Cohen, 2003), p. 101. It should also be noted that the methodology of this study was rather odd. Different toys were being presented to the infant during the experimental procedure, which could have differentially distracted some infants more than others. It's also noteworthy that what was measured was *frequency* of eye contact (actually, it was not even eye contact, but looking 'at the face region of the parent', p. 329) rather than *duration* of eye contact, although the two were correlated.

7 (Knickmeyer, Baron-Cohen et al., 2005). Multiple regression found that foetal testosterone predicted social relationships score independently of sex. However, within each sex no significant relationships were observed. It's also worth noting that the difference between boys and girls on this scale was not statistically significant (although there was a trend, with a moderate size of effect) and previous research with the same scale in six-year-olds found no sex differences. So even if amniotic testosterone does indeed correlate with the skills this questionnaire measures, there is not yet convincing evidence that males and females actually differ on them.

8 (Knickmeyer, Baron-Cohen et al., 2006). In this study, four-year-old children watched animations involving shapes. In two of the films, the behaviour of the shapes evokes the perception that they are acting on the basis of mental states. Children were interviewed about what was going on in the film. This involved extensive questioning by an interviewer (see p. 285). There is no mention of this interviewer being blind to experimental hypothesis or amniotic testosterone (aT) status, which seems problematic because an experimenter could unintentionally respond more encouragingly to girls, for example. Use of mental state (expressing character's beliefs, thoughts, intentions, etc.) and affective state terms (e.g., happy, sad) did not correlate with aT for all children, or within boys or girls. Although girls used significantly more affective state terms than boys, the sexes did not differ in mental state term use. For intentional propositions (e.g., 'the triangle knew the way'), aT was the only significant predictor in the hierarchical regression analysis. However, within females there was no correlation between aT and use of intentional propositions, but there was a correlation in males. The sex difference in intentional propositions use was at trend level. Boys used more neutral propositions than girls (e.g., 'There's a small triangle'). But although aT was the only significant predictor of neutral propositions, aT did not correlate with neutral propositions within boys and girls separately. All in all, the number of negative findings do not make for compelling evidence for the thesis that aT

levels are related to the tendency to attribute mental states to animated shapes, and that this tendency reliably differs in males and females.

9 (Chapman et al., 2006) For the Empathy Quotient (EQ)–child version, the only significant predictor in the hierarchical regression analysis was sex. In other words, amniotic testosterone was unrelated to EQ and something other than amniotic testosterone accounts for the effect of sex on score. There was a within-sex negative correlation between amniotic testosterone and EQ score for boys but not girls.

10 For the Reading the Mind in the Eyes test–child version, data confirmed hypotheses. However, as noted in the text, performance did not significantly differ – indeed, the authors report that they have previously failed to find superior performance of girls on this task (Chapman et al., 2006), see p. 140. This, in itself, seems a bit problematic for Baron-Cohen's thesis. A sex difference in a performance measure would be more convincing than maternal reports of sex differences.

11 Recently, Auyeung et al. (2009) reported correlations between amniotic testosterone and subclinical autistic traits, using two questionnaires. One of the questionnaires, the Autism Spectrum Quotient Child, was separable into subcomponents that included a mind-reading scale and a social skills scale. However, although these subscales both correlated with foetal testosterone, the authors do not present within-sex correlations.

12 (Voracek & Dressler, 2006), for example, found no relationship between digit ratio and either EQ score or Reading the Mind in the Eyes performance, in their large-scale study. As noted earlier, however, I do not attempt here to review the digit-ratio findings.

13 (Auyeung et al., 2006), p. S124.

14 (Levy, 2004), p. 319 citing Einstein quotations from H. L. Dreyfus & S. E. Dreyfus, *Mind over Machine* (New York: Macmillan, 1988), p. 41.

15 (Baron-Cohen, 2007), p. 161.

16 (Marton, Fensham, & Chaiklin, 1994). Both quotations on p. 467, from Yuan T. Lee and Konrad Lorenz.

17 (Houck, 2009), p. 66.

18 (Auyeung et al., 2006).

19 Baron-Cohen argues that systemising 'needs an exact eye for detail, since it makes a world of difference if you confuse one input or operation for another.' (Baron-Cohen, 2003), p. 64. However, it seems to me that one could just as plausibly argue that good empathising requires attention to detail, because otherwise you might, for example, fail to notice the important emotional leak that tells you what the other person is *really* feeling, or how you might be best able to make him or her feel better. In addition, the benefit of attention to detail

would seem to depend on whether the right detail is being attended to. Focus on something irrelevant will not be helpful to understanding a system. And sometimes, as the earlier quotations from the Nobel Prize winners suggest, breakthroughs in understanding require a feel for the bigger picture, beyond the details of the component parts.

20 (van de Beek et al., 2009). There was an unexpected positive correlation between levels of amniotic progesterone (a hormone associated more strongly with females) and playing with boyish toys! The researchers suggest that this may be a spurious effect.

21 Speed of rotation did correlate positively in girls with amniotic testosterone (aT), but boys' rotation speed seemed to get *slower* with increasing aT, and they performed no better than the girls (Grimshaw, Sitarenios, & Finegan, 1995). And as Hines points out, it is performance accuracy – which did not relate to aT – on which a sex difference is normally seen (Hines, 2006a).

22 (Finegan, Niccols, & Sitarenios, 1992). No sex differences in performance were seen.

23 (Auyeung, Baron-Cohen, Ashwin, Knickmeyer, Taylor, & Hackett, 2009), the Block Design Test. No sex difference in performance was seen.

24 (Brosnan, 2006; Puts et al., 2008; Voracek & Dressler, 2006).

25 (Gurian Institute, Bering, & Goldberg, 2009), p. 35.

26 (Connellan et al., 2000).

27 (Sax, 2006), p. 19.

28 (Lawrence, 2006), p. 15.

29 (Baron-Cohen, 2007), p. 169.

30 (Nash & Grossi, 2007).

31 (Nash & Grossi, 2007), p. 9.

32 (Leeb & Rejskind, 2004), pp. 4 and 10, respectively.

33 The article itself states that '[c]are was taken not to film any information that might indicate the sex of the baby' (p. 115), suggesting that such information was available. Additionally, in an interview with *Edge* magazine, Simon Baron-Cohen notes that sometimes Connellan did learn the sex of the baby because of clues such as congratulation cards (Edge, 2005a).

34 For example (Batki et al., 2000; Farroni et al., 2002). Regarding preference for motion, Philippe Rochat writes that '[i]nfants from birth tend to be more attentive to objects that move than to stationary objects. In devising experiments, researchers know that infants are much more engaged by dynamic compared to static displays.' (Rochat, 2001), p. 107. The study looking at preference for eye gaze (versus eyes closed) in newborns was conducted by the same team as Connellan's study, and may have used the same populations of newborns. (Connellan's face

was used as the stimulus for both studies.) Interestingly, this study found that newborn boys had no less of a preference for eye gaze than did girls.

35 (Nash & Grossi, 2007; Spelke, 2005). Spelke also highlights the lack of evidence that there are any sex differences in the acquisition of what she argues are the core cognitive systems that underlie mathematical ability.

36 A study of 119 same-sex three-year-old twins found no gender differences in a battery of Theory of Mind tasks (Hughes & Cutting, 1999) although a follow-up study with five-year-olds found a small advantage for girls (Hughes et al., 2005). This is consistent with a large body of research on young children's Theory of Mind skills, as noted by Nash and Grossi as well as development psychologist Alison Gopnik (Edge, 2005a). A meta-analysis of facial expression processing in children concluded that there is a small advantage for females (McClure, 2000). Yet it's not clear what we should make of this given that, as discussed in Chapter 2, men and women perform equivalently on the superior empathic accuracy task developed by William Ickes and colleagues. For meta-analysis of prosocial behaviour and empathic concern, see (Fabes & Eisenberg, 1998). Although Baron-Cohen argues that the rough-and-tumble play and direct (i.e., physical) aggression seen more commonly in males than females may reflect males' lower levels of empathy ('Direct aggression may require an even lower level of empathy than indirect aggression [such as spreading rumours, gossiping, and exclusion]'; [Baron-Cohen, 2007], p. 164), it is not clear that this is the case. One could, for example, argue that *successful* rough-and-tumble play demands quite high sensitivity to cues from one's play partner. Moreover, some research (although not all) finds that children find indirect aggression more harmful and hurtful than direct aggression (see discussion in [Archer & Coyne, 2005]).

37 (Levy, 2004), p. 322.

38 In addition to previously cited claims by Baron-Cohen regarding the implications for the gender gap in maths and physics, Connellan et al. claim that their findings 'demonstrate beyond reasonable doubt that [gender differences in sociability] are, in part, biological in origin.' (Connellan et al., 2000), p. 114. In my view, the methodology – as well as the undemonstrated link between newborn visual preferences and later sociability – allow ample room for extremely reasonable doubt.

39 (Baron-Cohen, 2007), p. 160.

11. THE BRAIN OF A BOY IN THE BODY OF A GIRL . . . OR A MONKEY?

1 Quoted in (Verghis, 2009), p. 26.

2 (Hoff Sommers, 2008), para. 31.

3 See (Houck, 2009).

4 (Schaffer, 2008), entry 6 ('The next best-seller'), para. 6.

5 For example (Hines, 2006a; Tavris, 1992, p. 54).

6 Gender identity in females with CAH seems to differ, albeit modestly, from control females. See for example (Berenbaum & Bailey, 2003), who found that gender identity scores of forty-three girls with CAH were intermediate between those of tomboys and sister controls, although this was not related to degree of genital virilisation or age of genital reconstructive surgery. A retrospective study of women with CAH found that women with the severest form of CAH had significantly greater cross-gender desire compared with controls (Meyer-Bahlburg et al., 2006). Also see (Hines, 2006b), figure 1, p. S117. Note that by gender identity I mean here responses to questions like, 'Do you ever wish you could be a boy?' rather than *confusion* over gender identity.

7 (Knickmeyer, Baron-Cohen, Fane et al., 2006; Mathews et al., 2009).

8 (Knickmeyer, Baron-Cohen, Fane et al., 2006).

9 See (Hines, 2004), p. 168.

10 (Puts et al., 2008).

11 (Pasterski et al., 2005).

12 For example (Berenbaum & Hines, 1992; Nordenström et al., 2002; Pasterski et al., 2005; Servin et al., 2003).

13 (Berenbaum, 1999). Also, (Servin et al., 2003) found stronger preference for masculine than feminine careers in seven- to ten-year-old girls with CAH, compared with controls.

14 It's been suggested, for example, that prenatal androgen levels function as 'the seeds of career choices' (Berenbaum & Resnick, 2007).

15 As Bleier pointed out in her critique of earlier studies in this area, 'authors and subsequent scientists accept at face value the idea of tomboyism [such as play preferences, clothing preferences, career interests, and so on] as an index of a characteristic called *masculinity*, presumed to be as objective and innate a human feature as height and eye colour. Yet 'masculinity' is a gender characteristic and, as such, culturally, not biologically, constructed' (Bleier, 1986), p. 150.

16 (Golombok & Rust, 1993).

17 As found by (Hines et al., 2003).

18 (Jürgensen et al., 2007). The clinical population in this study had a 46,XY karyotype with a condition causing either partial or complete androgen insensitivity.

19 (Meyer-Bahlburg et al., 2006).

20 A Lego aeroplane had to be substituted for the Lincoln Logs in the UK sample because it didn't show the expected sex difference in the US sample (Pasterski et al., 2005). Along similar lines, an earlier study found that control girls played

with Lincoln Logs more than *any other toy*, masculine or feminine (Servin et al., 2003). While it's hardly the most scientific of sources, data from the Fat Brain Toys Web site suggest that parents and others underestimate how much girls will enjoy Lincoln Logs. The vast majority of these products (roughly 80 percent when I looked) are bought for boys.

21 (Berenbaum, 1999; Jürgensen et al., 2007; Meyer-Bahlburg et al., 2004).

22 (Auyeung, Baron-Cohen, Ashwin, Knickmeyer, Taylor, Hackett et al., 2009; Hines et al., 2002). (Udry, 2000) found a relationship between maternal levels of SHBG (which, as it binds to testosterone, can be understood as an inverse measure of free testosterone − see note 3 in chapter 10) and adult gendered behaviour. As noted in the earlier footnote, it is unclear whether mT or SHBG or neither is the appropriate index of exposure of the foetus to androgens. It's also hard to determine from the information provided in this study to what extent the gendered behaviours measured indexed cultural ascriptions versus behaviours that are more plausibly regarded as psychological predispositions. (Knickmeyer, Wheelwright et al., 2005) found no relationship between aT and gender-typed play.

23 (Berenbaum, 1999), p. 108.

24 (Burton, 1977).

25 (Hines, 2004), pp. 127 and 128.

26 (Alexander & Hines, 2002). This study design, by the way, introduces other factors that might influence why a monkey might spend longer with a ball on Monday than a doll on Tuesday. For example, something that, to a monkey, is incredibly interesting might be taking place in the enclosure on Monday, while on Tuesday he may simply be in a less playful mood.

27 Frances Burton, personal communication, July 21, 2009. The study authors suggest that the appeal of the pan to the female monkeys may have been due to its red colour.

28 As noted by Ian Gold, Frances Burton and Lesley Rogers in their personal communication with me.

29 (Hassett, Siebert, & Wallen, 2008), p. 361. Although the researchers recorded the type of interaction with the toys, these data are not presented. The results are slightly different depending on whether total frequency or total duration of interaction is used. In the former case, the contrast between male and female plush toy play is also significant.

30 See (Hines & Alexander, 2008).

31 (Hines & Alexander, 2008), p. 478.

32 (Hassett et al., 2008), p. 363.

33 (Sax, 2006), p. 28.

34 (Mathews et al., 2009), replicating an earlier study discussed by Anne Fausto-Sterling. She points out that the idea that high foetal testosterone reduces

interest in infants implies 'that testosterone interferes with the development of interest in infants, but that some general character called nurturance, which could get directed everywhere but to children, existed independently of high androgen levels.' (Fausto-Sterling, 2000), pp. 289 and 290.

35 (Herman, Measday, & Wallen, 2003), p. 582. It should be noted that the findings with this androgen receptor blocker are sometimes paradoxical, suggesting that it may not have a straightforward androgen-blocking effect. However, early in gestation it does have the expected feminising effect on genitalia.

36 (Burton, 1977).

37 (Itani, 1959), p. 61.

38 (Burton, 1992), p. 45.

39 (Burton, 1977), pp. 11 and 14.

40 (Mason, 2002), p. 124.

41 (Herman, Measday, & Wallen, 2003). This study found that at one year of age, females differed from males only in touch behaviour, that is, the animal briefly touches the infant with its hand, although overall infant interaction approached significance.

42 See (Itani, 1959).

43 (Burton, 1977), p. 11.

44 (Burton, 1972).

45 (Hines & Alexander, 2008), p. 479.

46 (Hines, 2004), p. 181.

47 (Hines, 2004), p. 178.

48 Quoted in (Edge, 2005b).

49 (Pinker, 2005), para. 7.

50 (Baron-Cohen, 2005).

51 (Kimura, 2005), para. 2.

52 (Pinker, 2005), para. 12.

12. SEX AND PREMATURE SPECULATION

1 (Dana, 1915), para. 8.

2 (Russett, 1989), p. 191.

3 This section summarised from (Russett, 1989); quotation from p. 32. See also (Shields, 1975; Tavris, 1992).

4 (Hines, 2004), p. 6.

5 (Pease & Pease, 2008), p. 51.

6 Geoffrey Aguirre, quoted in (Lehrer, 2008), para. 17.

7 See (Weisberg, 2008) for an excellent overview.

8 See (Wallentin, 2009), also (Dietrich et al., 2001).

9 (Harrington & Farias, 2008; Ihnen et al., 2009; Kaiser et al., 2009). See also (Kriegeskorte et al., 2009; Vul et al., 2009) for arguments that reported correlations between brain activations and stimuli or social characteristics are sometimes biased or spurious due to invalid methods of analysis. Concern has also been expressed that the technology is being used in inappropriate ways. Neuroimaging expert Logothetis has recently complained that '[m]any of these [fMRI] papers are such oversimplifications of what's happening in the brain as to be worthless' and that '[t]oo many of these experiments are being done by people who, unfortunately, don't really understand what the technology can and cannot do.' (Quoted in [Lehrer, 2008], paras. 11 and 8, respectively.)

10 As Bleier points out, there was no a priori reason to suggest that greater lateralisation would be associated with superior visuospatial abilities. She also provides a good critique of the original corpus callosum data and interpretation (Bleier, 1986).

11 (Bleier, 1986), p. 154. Bleier provides an excellent and concise summary of the issues with the greater male lateralisation hypothesis and the inadequacy of the data for it. See also (Kaplan & Rogers, 1994).

12 (Sommer et al., 2004; Sommer et al., 2008), p. 1850 of 2004 paper. For the role of publication bias in the investigation of sex differences in language lateralisation, see also (Kaiser et al., 2009).

13 When Sommer and colleagues looked separately at the different types of dichotic listening tasks used, they found that one type of task, called the CV(C) task, did yield the expected sex difference. Interestingly, the CV(C) was used exclusively by researchers interested in sex difference issues. (In fact, generally, studies that were specifically interested in sex differences tended to find them, whereas studies that merely mentioned sex in passing tended not to.) Suspecting publication bias, they looked for evidence of sex differences in lateralisation in the CV(C) in a huge data set called the Bergen Dichotic Listening Database. This is an unpublished data set that is three times larger than all the CV(C) studies from the meta-analysis combined. There were no sex differences.

14 (Mathews et al., 2004).

15 See (Wallentin, 2009).

16 The aphasia rate following right-hemisphere damage was 2 percent for men and 1 percent for women (D. Kimura, 'Sex differences in cerebral organisation for speech and praxic functions', *Canadian Journal of Psychology* 37 [1983], pp. 19–35), cited in (Sommer et al., 2004), p. 1849.

17 See (Hyde, 2005). Summarising the findings relating to language and communication from Hyde's meta-analysis, Cameron writes, '[i]n almost every case, the overall difference made by gender is either small or close to zero.

Two items, spelling accuracy and frequency of smiling, show a larger effect – but it is still only moderate, not large.' (Cameron, 2007), p. 43. Wallentin also concludes his review as follows: 'A small but consistent female advantage is found in early language development. But this seems to disappear during childhood. In adults, sex differences in verbal abilities, and in brain structure and function related to language processing are not readily identified.' (Wallentin, 2009), p. 181. Wallentin later draws attention to the file-drawer problem for research into sex difference in language skills.

18 See Bleier's discussion of the initial report in 1982 by De Lacoste-Utamsing and Holloway (C. De Lacoste-Utamsing & R. L. Holloway, 'Sexual dimorphism in the human corpus callosum', *Science*, 216 [1982]: 1413–1432) which was based on fourteen brains, of unknown age or cause of death, and obtained a result that did not reach statistical significance. Bleier also made the important points that it is not known whether the size of the corpus callosum is related to the number of fibres *or* whether the number of fibres is related to degree of lateralisation of hemispheric function *or* whether lateralisation of hemispheric function is related to visuospatial ability (Bleier, 1986).

19 (Fausto-Sterling, 2000), and (Bishop & Wahlsten, 1997), p. 581.

20 (Wallentin, 2009), p. 178.

21 For example, one study found similar lateralisation (right) activity in the superior parietal lobe in both men and women – with males outperforming females (Halari et al., 2006). Another found no sex difference in behaviour, and found that males showed more bilateral activation in the parietal lobe while females showed more right lateralisation in this region (Clements et al., 2006). Gur and colleagues, on the other hand, found increased right lateralisation in men, who outperformed women, in the inferior parietal region (Gur et al., 2000). Another study found no differences in performance and no differences in lateralisation (Dietrich et al., 2001). This study also found much greater brain activations in women during their high-oestrogen phase which hints at an interesting problem for gender difference research in this area. Other researchers matched male and female performance and found sex differences in activations (which didn't clearly suggest greater lateralisation in either group) that they suggested were due to different strategies in women and men (Jordan et al., 2002). Another study found no gender difference in either performance on brain activations, but significant brain activation differences between good and poor performers on the task (Unterrainer et al., 2000).

22 (Halpern et al., 2007), pp. 29 and 30.

23 (Baron-Cohen et al., 2005), p. 820, references removed.

24 Quoted in (Healy, 2006a), para. 14.

25 Quoted in (Healy, 2006b), para .22.

26 (Gurian & Stevens, 2004), p. 23.

27 (Pease & Pease, 2008), p. 110.

28 (Gray, 2008), see p. 39.

29 A point made by (Bleier, 1986).

13. WHAT DOES IT ALL MEAN, ANYWAY?

1 (Romanes, 1887/1987), p. 11, footnote removed.

2 (Fausto-Sterling, 1985), p. 260.

3 (De Vries, 2004), p. 1064.

4 An example of this, in the rat, is described by (Moore, 1995), p. 53.

5 (Moore, 1995), pp. 53 and 54. Similarly, Haier and colleagues have suggested that 'different brain designs may manifest equivalent intellectual performance.' (Haier et al., 2005), p. 320.

6 See (Im et al., 2008).

7 (Leonard et al., 2008), p. 2929.

8 (Im et al., 2008; Leonard et al., 2008). Leonard et al. quoted on p. 2929. Effects of sex were very small, or nonexistent, once effect of total brain volume was taken into account. Leonard et al.'s findings with regard to grey matter in proportion to total brain volume are consistent, too, with work by Luders and colleagues, who also conclude that 'brain size is the main variable determining the proportion of grey matter.' (Luders, Steinmetz, & Jancke, 2002), p. 2371. Im and colleagues also argue that their results show 'that sex effects are mostly explained by brain size effects in the cortical structure of human brains.' (Im et al., 2008), p. 2188.

9 (Giedd et al., 2006), p. 159.

10 (Fine, 1990), p. 133.

11 (Baron-Cohen, Knickmeyer, & Belmonte, 2005), p. 821.

12 (Gur & Gur, 2007), p. 196.

13 Ian Gold, personal communication, October 24, 2008.

14 I am very grateful to Ian Gold, whose insights have greatly enhanced my understanding of the problems inherent in trying to relate brain structure to brain function.

15 (Halari et al., 2006), see pp. 1 and 3.

16 (Gur et al., 1999).

17 Quoted in (University of Pennsylvania Medical Center, 1999), para. 7.

18 (Gur et al., 1999), p. 4071. Regarding the point that correlation doesn't mean

causation, some third factor (or complex of factors), like education, could enhance both white matter volume *and* spatial ability.

19 (Gur & Gur, 2007), p. 196.

20 (Gur et al., 2000), p. 166.

21 (O'Boyle, 2005; O'Boyle et al., 2005; Singh & O'Boyle, 2004).

22 Again, this is an issue raised long ago by Ruth Bleier who pointed out the circularity of the reasoning that men are superior in visuospatial skills because they have right-hemisphere lateralisation for visuospatial processing, and that right-hemisphere lateralisation is superior for visuospatial processing because men are superior at visuospatial processing and they show right-hemisphere lateralisation (Bleier, 1986).

23 See (Russett, 1989; Shields, 1975).

24 H. Ellis, *Man and Woman: A Study of Human Secondary Sexual Characteristics* (London: Walter Scott, 1894), p. 28. Quoted in (Russett, 1989), pp. 184 and 185.

25 (Pease & Pease, 2008), pp. 145 and 146, respectively. Illustrations appear on p. 145.

26 (Pease & Pease, 2008), p. 145.

27 The first study is C. M. McCormick, S. F. Witelson, and E. Kingstone, 'Left-handedness in homosexual men and women: Neuroendocrine implications', *Psychoneuroendocrinology* 15, no. 1 (1990), pp. 69–76. The second study is S. F. Witelson, 'The brain connection: The corpus callosum is larger in left-handers', *Science* 229, no. 4714 (1985), pp. 665–668.

28 (Hall et al., 2004). Although the Peases also describe the Witelson emotion study in the 1999 edition of their book, researchers often present their results before publication, which can take many years. I contacted Pease International in the hope that the Peases might be able to clarify to what research they are referring in this passage, but they were unable to assist.

29 (Pinker, 2008), p. 116.

30 In discussing these results, I focus on between-group comparisons between males and females, rather than within-group contrasts, on the basis of the argument made by Kaiser and colleagues that '[o]nly by comparing women and men directly with one another within one statistical test can significance be ensured.' (Kaiser et al., 2009), p. 54.

31 (Hall et al., 2004), p. 223.

32 (Hall et al., 2004), p. 223.

33 (Bennett et al., 2009), p. S125.

34 See (Ihnen et al., 2009).

35 For discussions of the role of reverse inferences in understanding cognitive mechanisms, limitations and conditions in which they are more or less likely to be a valid form of inference, see (Poldrack, 2006; Poldrack & Wagner, 2004).

36 For example (Blakemore et al., 2007; Burnett et al., 2009; Haier et al., 1992).

37 (Bird et al., 2004), p. 925.

38 (Buracas, Fine, & Boynton, 2005).

39 (Friston & Price, 2001), p. 275.

40 (Lehrer, 2008), para. 7.

41 (Kaiser et al., 2009).

42 (Miller, 2008), p. 1413.

43 Men's brains are, on average, about 8 to 10 percent larger than female brains. Beyond this, as Kaiser et al. have pointed out, results demonstrating sex differences in 'a/symmetries between the left and right hemisphere in anatomy and function, the size of the *corpus callosum*, and the *extent* of defined brain areas . . . have never been both conclusive and unchallenged' (Kaiser et al., 2009), p. 50, emphases in original, references removed. Also, as discussed in this chapter, what appear to be sex differences in brain structure may turn out to be differences between people with larger versus smaller brains. Nor does the existence of differences in the brain indicate their origins. One last point is the importance of not assuming that sex differences observed in the rat apply to humans. With these extremely important caveats in mind, a brief overview of research finding sex differences in brain anatomy, neurochemistry and function, and discussion of their potential importance in understanding clinical disorders, is provided in (Cahill, 2006).

44 (Weisberg, 2008), p. 56.

14. BRAIN SCAMS

1 (Gray, 2008), pp. 44 and 45, respectively.

2 (Gurian & Annis, 2008), p. 9.

3 (Gurian, 2003), p. 88.

4 (Gurian & Annis, 2008), p. 34.

5 (Gurian & Annis, 2008), p. 59, emphasis in original.

6 (Rogers, Zucca, & Vallortigara, 2004). Thanks to Lesley Rogers for alerting me to this study.

7 (Young & Balaban, 2006), p. 634.

8 http://itre.cis.upenn.edu/~myl/languagelog/archives/003923.html, accessed on October 5, 2009.

9 The study cited is (Raingruber, 2001).

10 (Brizendine, 2007), p. 162.

11 (Hall, 1978; Hall, 1984; McClure, 2000).

12 (Brizendine, 2007), p. 162.

13 The study cited is (Oberman et al., 2005).

14 (Brizendine, 2007), p. 163.

15 The study cited is (Singer et al., 2004).

16 (Brizendine, 2007), p. 163.

17 The study cited is T. Iidaka, 'fMRI study of age related differences in the medial temporal lobe responses to emotional faces', Society for Neuroscience, New Orleans [*sic*, should be San Diego], 2001. The first author confirmed that the research presented at this conference was subsequently published in (Iidaka et al., 2002) and that, as in the published report, gender differences were not mentioned.

18 The study cited is (Zahn-Waxler, Klimes-Dougan, & Slattery, 2000), p. 458, emphasis in original.

19 (Brizendine, 2007), p. 163.

20 The study cited is (Singer et al., 2006).

21 (Brizendine, 2007), pp. 163 and 164. Note that the researchers actually interpret their empathy-related responses to the pain of another as being limited to the affective aspect of the pain response, rather than the sensory aspects of pain.

22 (Brizendine, 2007), p. 158. The citations are, in order discussed in current text: (Orzhekhovskaia, 2005); (Uddin et al., 2005); (Oberman et al., 2005); (Ohnishi et al., 2004); and L. M. Oberman, 'There may be a difference in male and female mirror neuron functioning', personal communication, 2005.

23 Lindsay M. Oberman, personal communication (with me), October 21, 2008.

24 (Brizendine, 2007), p. 210.

25 (Brizendine, 2007), pp. 188 and 189.

26 http://itre.cis.upenn.edu/~myl/languagelog/archives/004926.html, accessed March 3, 2010.

27 Quoted in (Weil, 2008), para. 14.

28 (Sax, 2006), pp. 106 and 107 and p. 106, respectively. The study Sax bases this claim on is described on pp. 29 and 30 of his book *Why Gender Matters*.

29 See (Freese & Amaral, 2009).

30 The study cited is (Killgore, Oki, & Yurgelun-Todd, 2001).

31 Although negative emotions conveyed in faces can be contagious, the children were not asked to try to induce a particular mood, and it was not the purpose of the experimental design to induce negative emotion in the children.

32 Brain activity was measured in two small parts of the brain bilaterally, in the amygdala and a region of the dorsolateral prefrontal cortex. For further critique of Sax's interpretation of this study, see Mark Liberman's discussion at http://itre.cis.upenn.edu/~myl/languagelog/archives/003284.html.

33 http://itre.cis.upenn.edu/~myl/languagelog/archives/003284.html, accessed September 2, 2009.

34 Sax cites one other study as support for his claim that in women brain activity associated with negative affect is 'mostly up in the cerebral cortex' whereas in men it is 'stuck down in the amygdala' (Sax, 2006), p. 29. This study (Schneider et al., 2000), involving thirteen men and thirteen women, found increased activity in the right amygdala in males but not females during induced sadness (but similar left amygdala activity during induced sadness, and similar amygdala activation in both hemispheres during induced happiness). Gender differences in cortical activations during induced sadness and happiness are not discussed. Sax also cites two other studies as evidence that emotions are processed differently in the sexes. Although he does not claim that these studies support the hypothesis that negative emotional experience is more subcortical in males and cortical in females, for the sake of completeness it is worth noting that these studies do not offer support for this idea. The first study (Killgore & Yurgelun-Todd, 2001) did not involve emotional experience but looked at amygdala activity in seven men and six women as they looked at fearful or happy faces (compared with the control condition of looking at a small circle). It did not look at brain activations in cortical regions. Amygdala response while looking at fearful faces was similar in the two sexes. When looking at happy faces, amygdala activation was lateralised to the right in men but not women – a lateralisation difference, rather than a difference in the engagement of the amygdala per se. Second, Sax cites a meta-analysis of functional imaging studies of emotion (Wager et al., 2003) as evidence that emotions are processed differently in the sexes. However, the conclusions of this study are not consistent with the idea that emotional experience is more subcortical in males and more cortical in women. The authors tentatively summarise the gender differences from their analysis as follows: 'Men tend to activate posterior sensory and association cortex, left inferior frontal cortex, and dorsal striatum more reliably than women, whereas women tend to activate medial frontal cortex, thalamus, and cerebellum more reliably' (p. 528). Translation: Men [cortical, cortical, cortical, subcortical] versus Women [cortical, subcortical, subcortical].

35 (Bachelard & Power, 2008), para. 46.

36 (Sax, 2006), p. 102 (boys) and p. 104 (girls). The term 'neurofallacy' coined by (Racine et al., 2005). For details of hippocampus-cortex connections, take your pick from the articles in the 2000 Special Issue of the journal *Hippocampus* entitled 'The nature of hippocampal-cortical interaction: Theoretical and experimental perspectives'.

37 See (Sax, 2006), pp. 100–101. Perhaps the most important reason that implications for maths education cannot be drawn from the cited neuroimaging study is that it did not involve maths, or even numbers. Rather, the task involved navigating out of a complex three-dimensional virtual maze. The

control condition involved looking at a frozen shot of the maze and making key presses in response to flickering rectangles. We can immediately see that this study will not tell us anything about the parts of the brain involved in mathematical processing. Even if the debate concerned whether single-sex classrooms are necessary for lessons in virtual maze navigation, this study would not help us much. More male activity was seen in the left hippocampus while women showed greater activation in right prefrontal and parietal areas, but this is in the context of 'great overlap' between the sexes in which regions were activated (Grön et al., 2000), p. 405. It's impossible to make useful inferences from these differences. What do we make of greater male activation of the left hippocampus given that the right was activated equally in the sexes? What is the significance of greater female activation of the superior parietal lobule on one side of the brain but not the other? It does not make sense to say that only females use the cerebral cortex and only males use the hippocampus while performing spatial navigation (and even less sense to make this claim for maths)! Moreover, we don't know whether more activation means 'better'. It could mean 'less efficient'. Were the differences due to performance differences rather than sex per se? (The men were significantly faster at getting out of the maze.) What cognitive role are these regions playing in the performance of the task? We have no idea – which makes developing educational strategies on the basis of these findings impossible. Discussing a similar claim about sex differences in maths processing made by a commentator on the BBC's 'Today' programme, the blogger known as Neurosceptic provides a useful explanation of some of the confusion behind such claims (see http://neuroskeptic.blogspot. com/2008/11/educational-neuro-nonsense-or-return-of.html, accessed on September 10, 2009).

38 http://itre.cis.upenn.edu/~myl/languagelog/archives/004618.html, accessed December 9, 2009.

39 Quoted in (Garner, 2008), para. 7.

40 (Bruer, 1997), p. 4.

41 (Clarke, 1873).

42 (Lewontin, 2000), p. 208.

43 Quoted in (Garner, 2008), para. 3.

44 http://neuroskeptic.blogspot.com/2008/11/educational-neuro-nonsense-or-return-of.html, accessed September 2, 2009.

15. THE 'SEDUCTIVE ALLURE' OF NEUROSCIENCE

1 (Sax, 2005), para. 8. In fairness to Sax, he is following the lead of the authors of the research paper on which this claim is made. They found different patterns of EEG waves (synchrony versus asynchrony) in children at rest, related these EEG patterns to complex psychological processes like language, mathematics and social cognition (which, recall, the children were not engaged in), and then suggested that their results 'have implications for gender differences in "readiness-to-learn"' – even though they report no gender differences in any of the cognitive abilities their EEG data were supposedly tapping (Hanlon, Thatcher, & Cline, 1999), p. 503.

2 From Sax's Web site: http://www.whygendermatters.com, accessed on December 9, 2009. More recently, the NASSPE Web site (see http://www.singlesexschools.org/research-brain.htm) has drawn on a structural imaging study (Lenroot et al., 2007) to further bolster this argument. This study found sex differences in the trajectory of volume changes in the brain across time, although many of these differences did not survive correction for total brain volume, which is greater in boys. In any case, the psychological implications of these findings are unknown. As the researchers put it: 'Differences in brain size between males and females should not be interpreted as implying any sort of functional advantage or disadvantage.' (p. 1072).

3 Quoted in (Dakss, 2005), para. 29.

4 (Hyde et al., 2008).

5 (Kemper, 1990), p. 13.

6 (Racine, Bar-Ilan, & Illes, 2005), p. 160.

7 (Gurian & Stevens, 2005), p. 42.

8 http://itre.cis.upenn.edu/~myl/languagelog/archives/003246.html, accessed on October 5, 2009.

9 http://www.jsmf.org/neuromill/chaff.htm#bn64, accessed on October 5, 2009.

10 (Weisberg et al., 2008). A similar favouring of findings attained from neuro-scientific methods was found by (Morton et al., 2006).

11 (McCabe & Castel, 2008).

12 (Weisberg, 2008), p. 54.

13 (Gurian, Henley, & Trueman, 2001), p. 45 and see p. 53.

14 (Brescoll & LaFrance, 2004; Coleman & Hong, 2008; Dar-Nimrod & Heine, 2006; Thoman et al., 2008).

15 (Dar-Nimrod & Heine, 2006), p. 435.

16 (Kimura, 1999), p. 8.

17 See also arguments made by Bleier with regard to scientists' responsibility for

the presentation of data in their writing (Bleier, 1986), and also (Bishop & Wahlsten, 1997).

18 (Weisberg, 2008), p. 55.

19 Hats off to the bloggers who regularly discuss these issues, in particular the tireless Mark Liberman.

16. UNRAVELLING HARDWIRING

1 For details, and contrast with maturational viewpoint, see (Westermann et al., 2007), in particular figure 4, p. 80. Also (Lickliter & Honeycutt, 2003; Mareschal et al., 2007).

2 (Wexler, 2006), pp. 3 and 4.

3 (Bleier, 1984), p. 52, footnote removed.

4 (Grossi, 2008).

5 (Shields, 1982), pp. 778 and 779. See also (Shields, 1975).

6 As Steven Pinker put it (Edge, 2005b).

7 For a history of the Greater Male Variability hypothesis see (Shields, 1982).

8 E. L. Thorndike, *Educational Psychology* (1910), p. 35. Quoted in (Hollingworth, 1914), p. 510.

9 (Summers, 2005), para. 4.

10 Quoted in (Edge, 2005b).

11 (Pinker, 2008), p. 13.

12 (Hollingworth, 1914). Wendy Johnson, Andrew Carothers, and Ian Deary published a reanalysis of these data in 2008. They concluded that males were *especially* variable at lower levels of IQ. They also noted that, with a ratio of about 2 boys to 1 girl at the very highest levels of intelligence, this did not go very far in explaining the much steeper ratios for high-level academic physical science, maths, and engineering positions (Johnson, Carothers, & Deary, 2008), p. 520.

13 (Grossi, 2008), p. 98.

14 (Feingold, 1994).

15 (Hyde et al., 2008).

16 (Guiso et al., 2008).

17 (Penner 2008; Machin & Pekkarinen 2008). These latter authors stress the strong pattern of greater male variability, but the boy/girl ratio (shown in parentheses) at the top 5 percent of maths ability was more-or-less equal in Indonesia (0.91), Thailand (0.92), Iceland (1.04) and the UK (1.08). Penner found greater female variability in the Netherlands, Germany and Lithuania. For useful discussion of these data, see (Hyde & Mertz, 2009).

18 (Andreescu et al., 2008), p. 1248.

19 See (Andreescu et al., 2008), p. 1248.

20 (Andreescu et al., 2008), p. 1251.

21 (Andreescu et al., 2008), p. 1252.

22 (Andreescu et al., 2008), pp. 1253 and 1254. See table 7, p. 1253.

23 (Summers, 2005), para. 4.

24 (Pinker, 2005), para. 3.

25 (Dweck, 2007), p. 49.

26 See (Blackwell, Trzesniewski, & Dweck, 2007; Dweck, 2007; Good, Aronson, & Inzlicht, 2003).

27 This has been surprisingly little discussed in the academic literature, but see (Chalfin, Murphy, & Karkazis, 2008; Fine, 2008).

28 (Morton et al., 2009), pp. 661 and 656 (reference removed), respectively.

29 This is thanks, in no small part, to books aimed at a general audience that have critiqued popular myths of gender. Recent examples of such efforts include (Barnett & Rivers, 2004; Cameron, 2007; Fausto-Sterling, 1985, 2000; Rogers, 1999; Tavris, 1992).

30 This is a point made in a general way by the instigators of the Critical Neuroscience project, which 'holds that while neuroscience potentially discloses facts about behaviour and its instantiation in the brain, the cultural context of science interacts with these knowledge claims, adds new meaning to them and influences the experience of the people to whom they pertain' (Choudhury, Nagel, & Slaby, 2009), p. 66, references removed.

17. PRECONCEPTIONS AND POSTCONCEPTIONS

1 (Kane, 2006b); epigraph and block quotation included.

2 (Summers, 2005), para. 5.

3 Quoted in (Edge, 2005b).

4 (Brizendine, 2007), p. 34.

5 (Sax, 2006), p. 28.

6 (Kane, 2009), p. 373.

7 (Rothman, 1988), p. 130.

8 (Smith, 2005), pp. 51 and 52, respectively.

9 (Nosek, 2007b), p. 184.

10 See (Greenwald et al., 2009).

11 (Gonzalez & Koestner, 2005), p. 407.

12 (Jost, Pelham, & Carvallo, 2002), p. 597.

13 Jost et al. found even stronger evidence of implicit paternalism in the

nontraditional sample (in which the mother's last name was not the same as the father's last name), however, they don't present the data for an analysis in which identical names were excluded.

14 (Jost, Pelham, & Carvallo, 2002), p. 588.

15 (Orenstein, 2000).

18. PARENTING WITH A HALF-CHANGED MIND

1 (Moon, Cooper, & Fifer, 1993).

2 (Quinn et al., 2002).

3 For example (Kelly et al., 2007).

4 (Hornik, Risenhoover, & Gunnar, 1987).

5 (Barrett, Campos, & Emde, 1996).

6 (Nash & Krawczyk, 1994). See also (Pomerleau et al., 1990), although this research did not find differences in the youngest age group.

7 (Clearfield & Nelson, 2006).

8 (Donovan, Taylor, & Leavitt, 2007).

9 (Mondschein, Adolph, & Tamis-LeMonda, 2000).

10 For example (Adams et al., 1995; Dunn, Bretherton, & Munn, 1987; Fivush, 1989; Leaper, Anderson, & Sanders, 1998).

11 Several researchers have suggested that implicit attitudes should be especially likely to predict more spontaneous and less controllable behaviours and judgements (e.g., [Strack & Deutsch, 2004], and this is consistent with some experimental work. However, a recent meta-analysis suggests that implicit measures are equally capable of predicting more readily controlled behaviours (Greenwald et al., 2009).

12 (Castelli, De Dea, & Nesdale, 2008), p. 1512.

13 See discussion in (Castelli, Zogmaister, & Tomelleri, 2009). Also (Aboud & Doyle, 1996).

14 (Castelli, Zogmaister, & Tomelleri, 2009).

15 (Tenenbaum & Leaper, 2002). A weak relationship was found in the middle school years. This was a meta-analysis, therefore gender attitudes were assessed in different ways in children and adults.

16. (Weitzman, Birns, & Friend, 1985), p. 897.

17 (Lytton & Romney, 1991).

18 (Freeman, 2007).

19 (Kane, 2006a), quotations from pp. 156, 157, 158, 161, and 161, respectively.

20 (Kane, 2006a), p. 172.

21 In part, this is probably because males are higher status than females, and so it is more acceptable to cross up than to cross down. But as well, there is a fear

that feminine interests in boys portends future psychological maladjustment and homosexuality (Martin, 1990; Martin, 2005; Sandnabba & Ahlberg, 1999).

22 (Kane, 2008).

23 (Orenstein, 2000), p. 4.

24 (Alexander, Wilcox, & Woods, 2009).

25 The 'hot potato' effect, whereby children find attractive novel toys less appealing when they are labelled as being for the other sex, was demonstrated in four- to five-year-old children by (Martin, Eisenbud, & Rose, 1995).

26 (van de Beek et al., 2009).

27 (Servin, Bohlin, & Berlin, 1999).

28 (Sax, 2006), p. 26.

29 The potential importance of familiarity in preference has been noted by (Zosuls et al., 2009), for example.

30 As found by (Servin, Bohlin, & Berlin, 1999).

31 As found by (Campbell et al., 2000).

32 (Bandura & Bussey, 2004), p. 696.

19. 'GENDER DETECTIVES'

1 'What color for your baby?' *Parents'* 14, no. 3 (March 1939), p. 98. Quoted in (Paoletti, 1997), p. 32.

2 (Hurlbert & Ling, 2007; Alexander, 2003).

3 (Lawson, 2007). Quotations from paras. 4, 5, 8, 8, and 10, respectively.

4 (Paoletti, 1997), pp. 30 and 31, respectively.

5 The salience of gender in the social world, and the active role played by the child in gender development that the salience and importance of gender motivates, has been highlighted by a number of researchers, for example (Arthur et al., 2008; Bem, 1983; Bigler & Liben, 2007; Martin & Halverson, 1981). The material that follows all draws on the insights of Gender Schema Theory and especially Developmental Intergroup Theory.

6 (Gelman, Taylor, & Naguyen, 2004).

7 (Levy & Haaf, 1994).

8 For example (Serbin, Poulin-Dubois, & Eichstedt, 2002), also (Poulin-Dubois et al., 2002), who found that knowledge was seen earlier in girls than in boys.

9 (Zosuls et al., 2009).

10 (Martin, Ruble, & Szkrybalo, 2002; Martin & Halverson, 1981).

11 (Martin & Ruble, 2004), p. 67.

12 (Ruble, Lurye, & Zosuls, 2008), p. 2.

13 (Martin & Ruble, 2004), p. 68.

14 Carol Martin, personal communication, September 9, 2009.
15 (Martin, Eisenbud, & Rose, 1995).

20. GENDER EDUCATION

1 (Bem, 1983), p. 611.
2 (Bem, 1989).
3 (Bem, 1983), p. 612.
4 Referring to the title of Sandra Bem's autobiography (Bem, 1998). At the end of the book the Bems' children, Jeremy and Emily, then in their early twenties, reflect on their childhood experiences. (In addition to trying to raise 'gender-aschematic' children, the Bems also wanted to raise their children in an antihomophobic and sex-positive way.) Both were grateful for what their unconventional rearing had done for them (said Jeremy, 'I get to be a complete person. That's what it comes down to') and were positive about the beliefs their parents had tried to convey to them in their unconventional fashion, although sometimes they disagreed with the details of the implementation. As Sandra Bem acknowledged, there were difficulties for the children in having gender removed as a legitimate source of identity, yet having to live in a culture that remains highly gendered. Both children also noted the difficulty of accepting elements or desires in themselves that were conventionally gendered (for example, to enjoy typically masculine interactions or take pride in being a 'manly man' in the case of Jeremy, or to want to be a pretty girl, in the case of Emily). The children ended up pursuing stereotypical interests – Jeremy mathematics and Emily the arts.
5 (Bem, 1983), p. 613.
6 (Rhode, 1997), p. 19.
7 (David, Grace, & Ryan, 2004), p. 142, reference removed, referring to work done by Kay Bussey and David Perry.
8 (Fagot, 1985), see table 3, p. 1102.
9 (Fagot, 1985; Lamb, Easterbrooks, & Holden, 1980; Lamb & Roopnarine, 1979).
10 (Serbin et al., 1979).
11 (Bannerjee & Lintern, 2000).
12 Here I rely on the description of Woodward's work provided in (Paechter, 2007). Unfortunately, I was unable to locate David Woodward's thesis (D. Woodward, 'Nursery class children's formation of gender perspectives', Unpublished MPhil thesis, Faculty of Education and Language Studies, Open University, 2003).

13 (Rudman & Glick, 2008), p. 178.

14 (Weitzman et al., 1972), p. 1141.

15 (Gooden & Gooden, 2001; Hamilton et al., 2006).

16 (Novell, 2004) and (Telford, 2003), p. 4.

17 (Rush & La Nauze, 2006).

18 (Turner-Bowker, 1996).

19 (Lamb & Brown, 2006), p. 158.

20 (Frawley, 2008). Quotation is from book blurb, see p. 294.

21 (Diekman & Murnen, 2004).

22 (Evans & Davies, 2000).

23 (Anderson & Hamilton, 2005), p. 149.

24 (Aubrey & Harrison, 2004; Barner, 1999; Leaper et al., 2002; Thompson & Zerbinos, 1995).

25 (Lamb & Brown, 2006), see pp. 64 and 65.

26 (Pike & Jennings, 2005) – 40 percent versus 76.9 percent. The same effect was not found for Harry Potter Lego, although this may have been because many more children had seen the original version of this ad.

27 (Weitzman et al., 1972), p. 1129.

28 (Lamb & Brown, 2006), see pp. 159 and 160. An exception, I think, is Gertrude McFuzz, but as Lamb and Brown note, this girl bird is 'fancy, vain, and jealous' (p. 160) – and she also, in the end, has to be rescued by a male character.

29 (Hamilton et al., 2006). (Tepper & Cassidy, 1999) found that females were underrepresented in titles, pictures, and central roles, but contrary to prediction found no differences in emotional language used by male and female characters. (Turner-Bowker, 1996) analysed thirty Caldecott winners and runners up from 1984–1994 and found underrepresentation of females in titles and pictures, although not central roles.

30 (DeLoache, Cassidy, & Carpenter, 1987).

31 (Black et al., 2009; Davis, 2003; Drees & Phye, 2001; Furnham, Abramsky, & Gunter, 1997; Sheldon, 2004).

32 (Götz, 2008).

33 (Smith & Cook, 2008). Both the TV and movie surveys also found that characters were predominantly Caucasian.

34 http://www.thegeenadavisinstitute.org/about_us.php, accessed on October 5, 2009.

35 See brief review in (Miller, Trautner, & Ruble, 2006).

36 (Rudman & Glick, 2008), p. 82.

37 (Fagot, Leinbach, & O'Boyle, 1992), p. 229, referring to work reported in (Leinbach, Hort, & Fagot, 1997).

38 (Leinbach, Hort, & Fagot, 1993).

39 Unpublished work cited in (David, Grace, & Ryan, 2004). Information on the age of the children in the study was provided by the first author (Barbara David, personal communication, June 25, 2009).

21. THE SELF-SOCIALISING CHILD

1 (Walker, 2008).

2 (Ruble, Lurye, & Zosuls, 2008), p. 2.

3 (Miller et al., 2009). Appearance was the most commonly used type of stereotype for girls among preschoolers, first, and fourth/fifth graders, although not kindergartners.

4 See summary in (Tajfel & Turner, 1986).

5 (Patterson & Bigler, 2006). See also (Bigler & Liben, 2007).

6 A point made by (Arthur et al., 2008) and (Bem, 1983), for example.

7 A point made by (Rudman & Glick, 2008), p. 73. Interestingly, when children are encouraged to categorise by age (that is, kids versus adults) rather than gender, the adjectives they used to describe boys and girls change (Sani et al., 2003).

8 (Ruble, Lurye, & Zosuls, 2008).

9 (Rudman & Glick, 2008), p. 60, referring to research conducted by (Leinbach, Hort, & Fagot, 1993).

10 Barbara Hort, personal communication, September 17, 2009.

11 Developmental psychologists have pointed out that there are often methodological difficulties with studies that fail to find a relationship between gender knowledge and gender preferences. See (Martin, Ruble, & Szkrybalo, 2002; Miller, Trautner, & Ruble, 2006).

12 (Zosuls et al., 2009).

13 See (Miller, Trautner, & Ruble, 2006), pp. 315 and 316.

14 (Bradbard & Endsley, 1983; Bradbard et al., 1986; Martin, Eisenbud, & Rose, 1995; Masters et al., 1979; Thompson, 1975).

15 (Green, Bigler, & Catherwood, 2004).

16 (Kane, 2006b).

17 (Trautner et al., 2005).

18 This is a joke, rather than a scientific fact.

EPILOGUE: AND S-T-R-E-T-C-H!

1 This event is described by (Morantz-Sanchez, 1985), pp. 306 and 307. Morantz-Sanchez points out that '[i]ronically, women physicians were saying much the same thing as Richard Cabot in their public pronouncements.' (p. 307).

2 (Baruch, 1915), quotations from paras. 3 and 4, then paras. 7 and 8, respectively.

3 (Dana, 1915), para. 9.

4 Reported in (Nowlan, 2006), para. 9.

5 Quoted in (Morantz-Sanchez, 1985), p. 306.

6 (Dana, 1915), para. 10.

7 (Gould, 1981), pp. 28 and 29.

8 For example (Kay et al., 2004; Lockwood & Kunda, 1997; Shah, 2003; Welnsteln, Przybylskl, & Ryan, 2009).

9 (Kimmel, 2008), p. 4.

10 (Sapolsky, 1997), para. 6.

11 (Kaiser et al., 2009), p. 9, citing the insight of (Fausto-Sterling, 2000). For evidence relating to neuroplasticity, see (Draganski et al., 2004; Maguire et al., 2000).

12 (Kaplan & Rogers, 2003), p. 74.

13 (Fausto-Sterling, 2000), p. 5.

14 For example (Krendl et al., 2008; Wraga et al., 2006). Also (Hausmann et al., 2009) who found that circulating testosterone levels were higher in men who performed cognitive tasks after gender-stereotype priming, compared with controls.

15 (Schweder & Sullivan, 1993), p. 498.

16 (Kimmel, 2008), p. 341.

17 (Bleier, 1986), p. 148.

18 (Kaplan & Rogers, 2003), p. 231.

19 (Davies, 1989), p. x.

20 (Senior, 2009).

BIBLIOGRAPHY

Aboud, F. E., & Doyle, A.-B. (1996). Parental and peer influences on children's racial attitudes. *International Journal of Intercultural Relations, 20*(3/4), 371–383.

Adams, G., Garcia, D. M., Purdie-Vaughns, V., & Steele, C. M. (2006). The detrimental effects of a suggestion of sexism in an instruction situation. *Journal of Experimental Social Psychology, 42*(5), 602–615.

Adams, S., Kuebli, J., Boyle, P. A., & Fivush, R. (1995). Gender differences in parent-child conversations about past emotions: A longitudinal investigation. *Sex Roles, 33*(5/6), 309–323.

Alexander, G. M. (2003). An evolutionary perspective of sex-typed toy preferences: Pink, blue, and the brain. *Archives of Sexual Behavior, 32*(1), 7–14.

Alexander, G. M., & Hines, M. (2002). Sex differences in response to children's toys in nonhuman primates (*Cercopithecus aethiops sabaeus*). *Evolution and Human Behavior, 23*(6), 467–479.

Alexander, G. M., Wilcox, T., & Woods, R. (2009). Sex differences in infants' visual interest in toys. *Archives of Sexual Behavior, 38*(3), 427–433.

Allen, K. (2009, July 26). Could crash spell doom for City's boys' club? *The Observer*, 3.

Ambady, N., Shih, M., Kim, A., & Pittinsky, T. L. (2001). Stereotype susceptibility in children: Effects of identity activation on quantitative performance. *Psychological Science, 12*(5), 385–390.

Ames, D. R., & Kammrath, L. K. (2004). Mind-reading and metacognition: Narcissism, not actual competence, predicts self-estimated ability. *Journal of Nonverbal Behavior, 28*(3), 187–209.

Anderson, D., & Hamilton, M. (2005). Gender role stereotyping of parents in children's picture books: The invisible father. *Sex Roles, 52*(3/4), 145–151.

Andreescu, T., Gallian, J. A., Kane, J. M., & Mertz, J. E. (2008). Cross-cultural analysis of students with exceptional talent in mathematical problem solving. *Notices of the American Mathematical Society, 55*(10), 1248–1260.

Archer, J., & Coyne, S. M. (2005). An integrated review of indirect, relational, and social aggression. *Personality & Social Psychology Review, 9*(3), 212–230.

Aronson, J., Lustina, M. J., Good, C., Keough, K., Steele, C. M., & Brown, J. (1999). When white men can't do math: Necessary and sufficient factors in stereotype threat. *Journal of Experimental Social Psychology, 35*(1), 29–46.

Arthur, A. E., Bigler, R. S., Liben, L. S., Gelman, S. A., & Ruble, D. N. (2008). Gender stereotyping and prejudice in young children: A developmental intergroup perspective. In S. R. Levy & M. Killen (eds.), *Intergroup attitudes and relations in childhood through adulthood* (pp. 66–86). Oxford: Oxford University Press.

Ashby, J. S., Ryan, M. K., & Haslam, S. A. (2007, March). Legal work and the glass cliff: Evidence that women are preferentially selected to lead problematic cases. *William & Mary Journal of Women and the Law*, 775–793.

Aubrey, J. S., & Harrison, K. (2004). The gender-role content of children's favorite television programs and its links to their gender-related perceptions. *Media Psychology, 6*(2), 111–146.

Auyeung, B., Baron-Cohen, S., Ashwin, E., Knickmeyer, R., Taylor, K., & Hackett, G. (2009). Fetal testosterone and autistic traits. *British Journal of Psychology, 100*, 1–22.

Auyeung, B., Baron-Cohen, S., Ashwin, E., Knickmeyer, R., Taylor, K., Hackett, G., et al. (2009). Fetal testosterone predicts sexually differentiated childhood behaviour in girls and in boys. *Psychological Science, 20*(2), 144–148.

Auyeung, B., Baron-Cohen, S., Chapman, E., Knickmeyer, R., Taylor, K., & Hackett, G. (2006). Foetal testosterone and the child systemizing quotient. *European Journal of Endocrinology, 155*, S123–S130.

Bachelard, M. & Power, L. (2008, November 9). The class divide. *Sunday Age*, 15.

Banaji, M. R., & Hardin, C. D. (1996). Automatic stereotyping. *Psychological Science, 7*(3), 136–141.

Banaji, M. R., Nosek, B. A., & Greenwald, A. G. (2004). No place for nostalgia in science: A response to Arkes and Tetlock. *Psychological Inquiry, 15*(4), 279–310.

Bandura, A., & Bussey, K. (2004). On broadening the cognitive, motivational, and sociostructural scope of theorizing about gender development and functioning: Comment on Martin, Ruble, and Szkrybalo (2002). *Psychological Bulletin, 130*(5), 691–701.

Bannerjee, R., & Lintern, V. (2000). Boys will be boys: The effect of social evaluation concerns on gender-typing. *Social Development, 9*(3), 397–408.

Bargh, J., & Williams, E. (2006). The automaticity of social life. *Current Directions in Psychological Science, 15*(1), 1–4.

Barner, M. R. (1999). Sex-role stereotyping in FCC-mandated children's educational television. *Journal of Broadcasting & Electronic Media, 43*(4), 551–564.

Barnett, R. & Rivers, C. (2004). *Same difference: How gender myths are hurting our relationships, our children, and our jobs.* New York: Basic Books.

Barnyard, K., & Lewis, R. (2009). *Corporate sex*ism: *The sex industry's infiltration of the modern workplace.* London: The Fawcett Society.

Baron-Cohen, S. (1997). *Mindblindness: An essay on autism and theory of mind.* Cambridge, MA: MIT Press.

———. (2003). *The essential difference: Men, women and the extreme male brain.* London: Allen Lane.

———. (2005, August 8). The male condition. *New York Times*, 15.

———. (2007). Sex differences in mind: Keeping science distinct from social policy. In S. Ceci & W. Williams (eds.), *Why aren't more women in science? Top researchers debate the evidence* (pp. 159–172). Washington, DC: American Psychological Association.

———. (2009). Autism test 'could hit maths skills'. From *BBC News*: http://news.bbc.co.uk/2/hi/health/7736196.stm. Accessed February 2, 2009.

Baron-Cohen, S., Knickmeyer, R. C., & Belmonte, M. K. (2005). Sex differences in the brain: Implications for explaining autism. *Science, 310*, 819–823.

Barres, B. (2006, July 13). Does gender matter? *Nature, 442*, 133–136.

Barrett, K. C., Campos, J. J., & Emde, R. (1996). Infants' use of conflicting emotion signals. *Cognition and Emotion, 10*(2), 113–135.

Baruch, S. (1915, July 4). Why women lack great originality. *New York Times*, 10.

Batki, A., Baron-Cohen, S., Wheelwright, S., Connellan, J., & Ahluwalia, J. (2000). Is there an innate eye gaze module? Evidence from human neonates. *Infant Behavior & Development, 23*(2), 223–229.

Beilock, S. L., Rydell, R. J., & McConnell, A. R. (2007). Stereotype threat and working memory: Mechanisms, alleviation, and spillover. *Journal of Experimental Psychology: General, 136*(2), 256–276.

Belkin, L. (2003, October 26). The opt-out revolution. *New York Times Magazine*, 42.

———. (2008, June 15). When mom and dad share it all. *New York Times Magazine*, 44.

Bem, S. L. (1983). Gender schema theory and its implications for child development: Raising gender-aschematic children in a gender-schematic society. *SIGNS: Journal of Women in Culture & Society, 8*(4), 598–616.

———. (1989). Genital knowledge and gender constancy in preschool children. *Child Development, 60*, 649–662.

———. (1998). *An unconventional family*. New Haven & London: Yale University Press.

Bennett, C. M., Baird, A. A., Miller, M. B., and Wolford, G. L. (2009). Neural correlates of interspecies perspective taking in the post-mortem Atlantic Salmon: An argument for multiple comparisons correction. [Paper presented at the Organization for Human Brain Mapping 2009 Annual Meeting, San Francisco, CA] *NeuroImage, 47*(51), S125.

Berenbaum, S. A. (1999). Effects of early androgens on sex-typed activities and interests in adolescents with congenital adrenal hyperplasia. *Hormones and Behavior, 35*, 102–110.

Berenbaum, S. A., & Bailey, J. (2003). Effects on gender identity of prenatal androgens and genital appearance: Evidence from girls with congenital adrenal hyperplasia. *The Journal of Clinical Endocrinology and Metabolism, 88*(3), 1102–1106.

Berenbaum, S. A., & Hines, M. (1992). Early androgens are related to childhood sex-typed toy preferences. *Psychological Science, 3*(3), 203–206.

Berenbaum, S. A., & Resnick, S. (2007). The seeds of career choices: Prenatal sex hormone effects on psychology sex differences. In S. Ceci & C. Williams (eds.), *Why aren't more women in science? Top researchers debate the evidence* (pp. 147–157). Washington, DC: American Psychological Association.

Biernat, M., & Kobrynowicz, D. (1997). Gender- and race-based standards of competence: Lower minimum standards but higher ability standards for devalued groups. *Journal of Personality & Social Psychology, 72*(3), 544–557.

Bigler, R. S., & Liben, L. S. (2007). Developmental intergroup theory: Explaining and reducing children's social stereotyping and prejudice. *Current Directions in Psychological Science, 16*(3), 162–166.

Bird, C. M., Castelli, F., Malik, O., Frith, U., & Husain, M. (2004). The impact of extensive medial frontal lobe damage on 'Theory of Mind' and cognition. *Brain, 127*, 914–928.

Bishop, K. M., & Wahlsten, D. (1997). Sex differences in the human corpus callosum: Myth or reality? *Neuroscience and Biobehavioral Reviews, 21*(5), 581–601.

Bittman, M., England, P., Sayer, L., Folbre, N., & Matheson, G. (2003). When does gender trump money? Bargaining and time in household work. *American Journal of Sociology, 109*(1), 186–214.

Black, K., Marola, J., Littman, A., Chrisler, J., & Neace, W. (2009). Gender and form of cereal box characters: Different medium, same disparity. *Sex Roles, 60*(11/12), 882–889.

Blackwell, L. S., Trzesniewski, K. H., & Dweck, C. S. (2007). Implicit theories of intelligence predict achievement across an adolescent transition: A longitudinal study and an intervention. *Child Development, 78*(1), 246–263.

Blair, R. J. R. (1996). Morality in the autistic child. *Journal of Autism and Developmental Disorders, 26*(5), 571–579.

Blakemore, S.-J., den Ouden, H., Choudhury, S., & Frith, C. (2007). Adolescent development of the neural circuitry for thinking about intentions. *Social Cognitive and Affective Neuroscience, 2*, 130–139.

Blanton, H., Crocker, J., & Miller, D. T. (2000). The effects of in-group versus out-group social comparison on self-esteem in the context of a negative stereotype. *Journal of Experimental Social Psychology, 36*, 519–530.

Bledsoe, A. T. (1856). *An essay on liberty and slavery.* Philadelphia: J. B. Lippincott & Co.

Bleier, R. (1984). *Science and gender: A critique of biology and its theories on women.* New York: Pergamon Press.

———. (1986). Sex differences research: Science or belief? In R. Bleier (ed.), *Feminist approaches to science* (pp. 147–164). New York: Pergamon Press.

Blum, L., & Frieze, C. (2005). The evolving culture of computing: Similarity is the difference. *Frontiers, 26*(1), 110–125.

Bolino, M. C., & Turnley, W. H. (2003). Counternormative impression management, likeability, and performance ratings: The use of intimidation in an organizational setting. *Journal of Organizational Behavior, 24*(2), 237–250.

Bowles, H. R., Babcock, L., & Lai, L. (2007). Social incentives for gender differences in the propensity to initiate negotiations: Sometimes it does hurt to ask. *Organizational Behavior and Human Decision Processes, 103*, 84–103.

Bradbard, M. R., & Endsley, R. C. (1983). The effects of sex-typed labeling on preschool children's information-seeking and retention. *Sex Roles, 9*(2), 247–260.

Bradbard, M. R., Martin, C. L., Endsley, R. C., & Halverson, C. F. (1986). Influence of sex stereotypes on children's exploration and memory: A competence versus performance distinction. *Developmental Psychology, 22*(4), 481–486.

Breedlove, S. M., Cooke, B. M., & Jordan, C. L. (1999). The orthodox view of brain sexual differentiation. *Brain, Behavior and Evolution, 54*(1), 8–14.

Brescoll, V., & LaFrance, M. (2004). The correlates and consequences of newspaper reports of research on sex differences. *Psychological Science, 15*(8), 515–520.

Brescoll, V. L., & Uhlmann, E. L. (2008). Can an angry woman get ahead? Status conferral, gender, and expression of emotion in the workplace. *Psychological Science, 19*(3), 268–275.

Brines, J. (1994). Economic dependency, gender, and the division of labor at home. *American Journal of Sociology, 100*(3), 652–688.

Brizendine, L. (2007). *The female brain.* London: Bantam Press.

Broad, J. (2002). *Women philosophers of the seventeenth century*. Cambridge: Cambridge University Press.

Broad, J., & Green, K. (2009). *A history of women's political thought in Europe, 1400–1700*. Cambridge: Cambridge University Press.

Brosnan, M. J. (2006). Digit ratio and faculty membership: Implications for the relationship between prenatal testosterone and academia. *British Journal of Psychology, 97*(4), 455–466.

Bruer, J. T. (1997). Education and the brain: A bridge too far. *Educational Researcher, 26*(8), 4–16.

Bryden, M. P., McManus, I. C., & Bulman-Fleming, M. B. (1994). Evaluating the empirical support for the Geschwind-Behan-Galaburda model of cerebral lateralization. *Brain and Cognition, 26*(2), 103–167.

Buracas, G. T., Fine, I., & Boynton, G. M. (2005). The relationship between task performance and functional magnetic resonance imaging response. *Journal of Neuroscience, 25*(12), 3023–3031.

Burnett, S., Bird, G., Moll, J., Frith, C., & Blakemore, S. J. (2009). Development during adolescence of the neural processing of social emotion. *Journal of Cognitive Neuroscience, 21*(9), 1736–1750.

Burton, F. D. (1972). The integration of biology and behavior in the socialization of *Macaca sylvana* of Gibraltar. In F. E. Poirier (ed.), *Primate socialization* (pp. 29–62). New York: Random House.

———. (1977). Ethology and the development of sex and gender identity in non-human primates. *Acta Biotheoretica, 26*(1), 1–18.

———. (1992). The social group as information unit: Cognitive behaviour, cultural processes. In F. D. Burton (ed.), *Social processes and mental abilities in non-human primates: Evidences from longitudinal field studies* (pp. 31–60). Lewiston/Queenston/Lampeter: The Edwin Mellen Press.

Butler, D., & Geis, F. L. (1990). Nonverbal affect responses to male and female leaders: Implications for leadership evaluations. *Journal of Personality and Social Psychology, 58*(1), 48–59.

Cadinu, M., Maass, A., Frigerio, S., Impagliazzo, L., & Latinotti, S. (2003). Stereotype threat: The effect of expectancy on performance. *European Journal of Social Psychology, 33*(2), 267–285.

Cadinu, M., Maass, A., Rosabianca, A., & Kiesner, J. (2005). Why do women underperform under stereotype threat? Evidence for the role of negative thinking. *Psychological Science, 16*(7), 572–578.

Cahill, L. (2006). Why sex matters for neuroscience. *Nature Reviews Neuroscience, 7*(6), 477–484.

Cameron, D. (2007). *The myth of Mars and Venus: Do men and women really speak different languages?* Oxford: Oxford University Press.

Campbell, A., Shirley, L., Heywood, C., & Crook, C. (2000). Infants' visual preferences for sex-congruent babies, children, toys and activities: A longitudinal study. *British Journal of Developmental Psychology, 18*, 479–498.

Castelli, L., De Dea, C., & Nesdale, D. (2008). Learning social attitudes: Children's sensitivity to the nonverbal behaviors of adult models during interracial interactions. *Personality and Social Psychology Bulletin, 34*(11), 1504–1513.

Castelli, L., Zogmaister, C., & Tomelleri, S. (2009). The transmission of racial attitudes within the family. *Developmental Psychology, 45*(2), 586–591.

Ceci, S. J., Williams, W. M., & Barnett, S. M. (2009). Women's underrepresentation in science: Sociocultural and biological considerations. *Psychological Bulletin, 135*(2), 218–261.

Chalfin, M. C., Murphy, E. R., & Karkazis, K. A. (2008). Women's neuroethics? Why sex matters for neuroethics. *American Journal of Bioethics, 8*(1), 1–2.

Chapman, E., Baron-Cohen, S., Auyeung, B., Knickmeyer, R., Taylor, K., & Hackett, G. (2006). Fetal testosterone and empathy: Evidence from the Empathy Quotient (EQ) and the 'Reading the Mind in the Eyes' test. *Social Neuroscience, 1*(2), 135–148.

Charles, M., & Bradley, K. (2009). Indulging our gendered selves? Sex segregation by field of study in 44 countries. *American Journal of Sociology, 114*(4), 924–976.

Chatard, A., Guimond, S., & Selimbegovic, L. (2007). 'How good are you in math?' The effect of gender stereotypes on students' recollection of their school marks. *Journal of Experimental Social Psychology, 43*(6), 1017–1024.

Cherney, I. D. (2008). Mom, let me play more computer games: They improve my mental rotation skills. *Sex Roles, 59*, 776–786.

Cheryan, S., Plaut, V. C., Davies, P. G., & Steele, C. M. (2009). Ambient belonging: How stereotypical cues impact gender participation in computer science. *Journal of Personality & Social Psychology, 97*(6), 1045–1060.

Choudhury, S., Nagel, S. K., & Slaby, J. (2009). Critical neuroscience: Linking neuroscience and society through critical practice. *BioSocieties, 4*, 61–77.

Clarke, E. H. (1873). *Sex in education: Or, a fair chance for girls.* Boston: James R. Osgood & Company. Released by The Project Gutenberg, June 5, 2006.

Clearfield, M. W., & Nelson, N. M. (2006). Sex differences in mothers' speech and play behavior with 6-, 9-, and 14-month-old infants. *Sex Roles, 54*(1/2), 127–137.

Clements, A. M., Rimrodt, S. L., Abel, J. R., Blankner, J. G., Mostofsky, S. H., Pekar, J. J., et al. (2006). Sex differences in cerebral laterality of language and visuospatial processing. *Brain and Language, 98*(2), 150–158.

Coleman, J. M., & Hong, Y. Y. (2008). Beyond nature and nurture: The influence of lay gender theories on self-stereotyping. *Self and Identity, 7*(1), 34–53.

Collins, G. (1982, May 17). New perspectives on father and his role. From the *New York Times* online: http://www.nytimes.com/1982/05/17/style/new-perspectives-on-father-and-his-role.html, accessed March 7, 2009.

Connellan, J., Baron-Cohen, S., Wheelwright, S., Batki, A., & Ahluwalia, J. (2000). Sex differences in human neonatal social perception. *Infant Behavior & Development, 23*, 113–118.

Correll, S. J. (2004). Constraints into preferences: Gender, status, and emerging career aspirations. *American Sociological Review, 69*(1), 93–113.

———. (2001). Gender and the career choice process: The role of biased self-assessment. *American Journal of Sociology, 106*(6), 1691–1730.

Correll, S. J., Benard, S., & Paik, I. (2007). Getting a job: Is there a motherhood penalty? *American Journal of Sociology, 112*(5), 1297–1338.

Costa, P., Jr., Terracciano, A., & McCrae, R. R. (2001). Gender differences in personality traits across cultures: Robust and surprising findings. *Journal of Personality & Social Psychology, 81*(2), 322–331.

Croizet, J.-C., Després, G., Gauzins, M.-E., Huguet, P., Leyens, J.-P., & Méot, A. (2004). Stereotype threat undermines intellectual performance by triggering a disruptive mental load. *Personality and Social Psychology Bulletin, 30*(6), 721–731.

Crosby, F. J., Williams, J. C., & Biernat, M. (2004). The maternal wall. *Journal of Social Issues, 60*(4), 675–682.

Cuddy, A. J. C., Fiske, S. T., & Glick, P. (2004). When professionals become mothers, warmth doesn't cut the ice. *Journal of Social Issues, 60*(4), 701–718.

Dakss, B. (2005, March 13). Intellectual gender gap? Kaledin on debate reignited by Harvard president's comments. From *CBS News.com*: http://www.cbsnews.com/stories/2005/03/14/sunday/main679829.shtml. Accessed on November 24, 2009.

Dana, C. L. (1915, June 27). Suffrage a cult of self and sex. *New York Times*, 14, (online archive).

Danaher, K., & Crandall, C. S. (2008). Stereotype threat in applied settings re-examined. *Journal of Applied Social Psychology, 38*(6), 1639–1655.

Dar-Nimrod, I., & Heine, S. J. (2006). Exposure to scientific theories affects women's math performance. *Science, 314*(5798), 435.

Dasgupta, N., & Asgari, S. (2004). Seeing is believing: Exposure to counterstereotypic women leaders and its effect on the malleability of automatic gender stereotyping. *Journal of Experimental Social Psychology, 40*, 642–658.

David, B., Grace, D., & Ryan, M. K. (2004). The gender wars: A self-categorisation theory perspective on the development of gender identity. In M. Bennett & F. Sani (eds.), *The development of the social self* (pp. 135–157). Hove, UK, and New York: Psychology Press.

Davies, B. (1989). *Frogs and snails and feminist tales: Preschool children and gender.* Sydney: Allen & Unwin.

Davies, P. G., Spencer, S. J., Quinn, D. M., & Gerhardstein, R. (2002). Consuming images: How television commercials that elicit stereotype threat can restrain women academically and professionally. *Personality and Social Psychology Bulletin, 28*(12), 1615–1628.

Davies, P. G., Spencer, S. J., & Steele, C. M. (2005). Clearing the air: Identity safety moderates the effects of stereotype threat on women's leadership aspirations. *Journal of Personality and Social Psychology, 88*(2), 276–287.

Davis, M. H., & Kraus, L. A. (1997). Personality and empathic accuracy. In W. J. Ickes (ed.), *Empathic accuracy* (pp. 144–168). New York: The Guilford Press.

Davis, S. N. (2003). Sex stereotypes in commercials targeted toward children: A content analysis. *Sociological Spectrum, 23*, 407–424.

———. (2007). Gender ideology construction from adolescence to young adulthood. *Social Science Research, 36*, 1021–1041.

Davison, H. K., & Burke, M. J. (2000). Sex discrimination in simulated employment contexts: A meta-analytic investigation. *Journal of Vocational Behavior, 56*(2), 225–248.

De Vries, G. J. (2004). Sex differences in adult and developing brains: Compensation, compensation, compensation. *Endocrinology, 145*(3), 1063–1068.

DeLoache, J. S., Cassidy, D. J., & Carpenter, C. J. (1987). The three bears are all boys: Mothers' gender labeling of neutral picture book characters. *Sex Roles, 17*(3/4), 163–178.

Demos, J. (1982). The changing faces of fatherhood: A new exploration in American family history. In S. Cath, A. Gurwitt, & J. Ross (eds.), *Father and child: Developmental and clinical perspectives* (pp. 425–445). Boston: Little, Brown.

Deutsch, F. (1999). *Halving it all: How equally shared parenting works.* Cambridge, MA: Harvard University Press.

Devos, T., Diaz, P., Viera, E., & Dunn, R. (2007). College education and motherhood as components of self-concept: Discrepancies between implicit and explicit assessments. *Self and Identity, 6*(2/3), 256–277.

Diamond, M. C. (1991). Hormonal effects on the development of cerebral lateralization. *Psychoneuroendocrinology, 16*(1–3), 121–129.

Diekman, A. B., & Murnen, S. K. (2004). Learning to be little women and little men: The inequitable gender equality of nonsexist children's literature. *Sex Roles, 50*(5/6), 373–385.

Dietrich, T., Krings, T., Neulen, J., Willmes, K., Erberich, S., Thron, A., & Sturm, W. (2001). Effects of blood estrogen level on cortical activation patterns during cognitive activation as measured by functional MRI. *NeuroImage, 13*(3), 425–432.

Donovan, W., Taylor, N., & Leavitt, L. (2007). Maternal sensory sensitivity and response bias in detecting change in infant facial expressions: Maternal self-efficacy and infant gender labeling. *Infant Behavior and Development, 30*(3), 436–452.

Dorr, R. C. (1915, September 19). Is woman biologically barred from success? *New York Times*, SM15 (online archive).

Draganski, B., Gaser, C., Busch, V., Schuierer, G., Bogdahn, U., & May, A. (2004). Neuroplasticity: Changes in grey matter induced by training. *Nature, 427*(6972), 311–312.

Drake, J. (1696). *An essay in defence of the female sex. In which are inserted the characters of a pedant, a squire, a vertuoso, a poetaster, a city-critick, &c. In a letter to a lady. Written by a lady.* London: Printed for A. Roper, E. Wilkinson, and R. Clavel.

Drees, D. D., & Phye, G. D. (2001). Gender representation in children's language arts computer software. *Journal of Educational Research, 95*(1), 49–55.

Dugan, E. (2008, March 31). Women's group calls for the end to city's lap-dancing culture. *The Independent*, 12.

Dunn, J., Bretherton, I., & Munn, P. (1987). Conversations about feeling states between mothers and their young children. *Developmental Psychology, 23*(1), 132–139.

Dweck, C. S. (2007). Is math a gift? Beliefs that put females at risk. In S. Ceci & W. Williams (eds.), *Why aren't more women in science? Top researchers debate the evidence* (pp. 47–55). Washington, DC: American Psychological Association.

Dziobek, I., Rogers, K., Fleck, S., Bahnemann, M., Heekeren, H. R., Wolf, O. T., & Convit, A. (2008). Dissociation of cognitive and emotional empathy in adults with Asperger syndrome using the Multifaceted Empathy Test (MET). *Journal of Autism and Developmental Disorders, 38*(3), 464–473.

Edge (2005a). The assortative mating theory: A talk with Simon Baron-Cohen. http://www.edge.org/3rd_culture/baron-cohen05/baron-cohen05_index.html. Accessed September 2, 2009.

Edge (2005b, May 16). The science of gender and science. Pinker vs. Spelke. A debate. http://www.edge.org/3rd_culture/debate05/debate05_index.html. Accessed January 20, 2009.

Eisenberg, N., & Lennon, R. (1983). Sex differences in empathy and related capacities. *Psychological Bulletin, 94*(1), 100–131.

Evans, L., & Davies, K. (2000). No sissy boys here: A content analysis of the representation of masculinity in elementary school reading textbooks. *Sex Roles, 42*(3/4), 255–270.

Fabes, R. A., & Eisenberg, N. (1998). Meta-analyses of age and sex differences in children's and adolescents' prosocial behavior. (Manuscript partially published in: Eisenberg, N., & Fabes, R. A. (1998). Prosocial development. In W. Damon (ed.), Handbook of Child Development. Available at http://www.public.asu.edu/~rafabes/meta.pdf.)

Fagot, B., Leinbach, M., & O'Boyle, C. (1992). Gender labeling, gender stereotyping, and parenting behaviors. *Developmental Psychology, 28*(2), 225–230.

Fagot, B. I. (1985). Beyond the reinforcement principle: Another step toward understanding sex role development. *Developmental Psychology, 21*(6), 1097–1104.

Fara, P. (2005). *Scientists anonymous: Great stories of women in science.* Cambridge: Icon Books.

Farroni, T., Csibra, G., Simion, F., & Johnson, M. H. (2002). Eye contact detection in humans from birth. *Proceedings of the National Academy of Sciences, USA, 99*(14), 9602–9605.

Fausto-Sterling, A. (1985). *Myths of gender: Biological theories about women and men* (2nd ed.). New York: Basic Books.

———. (2000). *Sexing the body: Gender politics and the construction of sexuality.* New York: Basic Books.

Feingold, A. (1994). Gender differences in variability in intellectual abilities: A cross-cultural perspective. *Sex Roles, 30*(1/2), 81–92.

Feng, J., Spence, I., & Pratt, J. (2007). Playing an action video game reduces gender differences in spatial cognition. *Psychological Science, 18*(10), 850–855.

Fine, A. (1990). *Taking the devil's advice.* London: Viking.

Fine, C. (2006). *A mind of its own: How your brain distorts and deceives.* New York: W. W. Norton.

———. (2008). Will working mothers' brains explode? The popular new genre of neurosexism. *Neuroethics, 1*(1), 69–72.

Finegan, J. K., Niccols, G. A., & Sitarenios, G. (1992). Relations between prenatal testosterone levels and cognitive abilities at 4 years. *Developmental Psychology, 28*(6), 1075–1089.

Fiske, S. T., Cuddy, A. J. C., & Glick, P. (2007). Universal dimensions of social cognition: Warmth and competence. *Trends in Cognitive Sciences, 11*(2), 77–83.

Fivush, R. (1989). Exploring sex differences in the emotional content of mother-child conversations about the past. *Sex Roles, 20*(11/12), 675–691.

Frawley, T. J. (2008). Gender schema and prejudicial recall: How children misremember, fabricate, and distort gendered picture book information. *Journal of Research in Childhood Education, 22*(3), 291–303.

Freeman, N. K. (2007). Preschoolers' perceptions of gender appropriate toys and their parents' beliefs about genderized behaviors: Miscommunication, mixed messages, or hidden truths? *Early Childhood Education Journal, 34*(5), 357–366.

Freese, J. L. & Amaral, D. G. (2009). Neuroanatomy of the primate amygdala. In P. J. Whalen & E. A. Phelps (eds.), *The human amygdala* (pp. 3–42). New York: Guilford Press.

Friston, K. J., & Price, C. J. (2001). Dynamic representations and generative models of brain function. *Brain Research Bulletin, 54*(3), 275–285.

Fullagar, C. J., Canan Sumer, H., Sverke, M., & Slick, R. (2003). Managerial sex-role stereotyping, a cross cultural analysis. *International Journal of Cross Cultural Management, 3*(1), 93–107.

Furnham, A., Abramsky, S., & Gunter, B. (1997). A cross-cultural content analysis of children's television advertisements. *Sex Roles, 37*(1/2), 91–99.

Galinsky, A. D., Wang, C. S., & Ku, G. (2008). Perspective-takers behave more stereotypically. *Journal of Personality and Social Psychology, 95*(2), 404–419.

Galpin, V. (2002). Women in computing around the world. *SIGCSE Bulletin, 34*(2), 94–100.

Garner, H. (2004). *Joe Cinque's consolation: A true story of death, grief and the law.* Sydney: Pan Macmillan.

Garner, R. (2008, November 18). Single-sex schools 'are the future'. From *The Independent* online: http://www.independent.co.uk/news/education/education-news/singlesex-schools-are-the-future-1023105.html, accessed on May 22, 2009.

Gawronski, B., & Bodenhausen, G. V. (2006). Associative and propositional processes in evaluation: An integrative review of implicit and explicit attitude on change. *Psychological Bulletin, 132*(5), 692–731.

Gelman, S. A., Taylor, M. G., & Naguyen, S. P. (2004). III. How children and mothers express gender essentialism. *Monographs of the Society for Research in Child Development, 69*(1), 33–63.

Geschwind, N., & Behan, P. (1982). Left-handedness: Association with immune disease, migraine, and developmental learning disorder. *Proceedings of the National Academy of Sciences, USA, 79,* 5097–5100.

Gharibyan, H. (2007). Gender gap in computer science: Studying its absence in one former Soviet Republic. 2007 American Society for Engineering Education Conference Proceedings, Honolulu, Hawaii.

Gharibyan, H., & Gunsaulus, S. (2006). Gender gap in computer science does not exist in one former Soviet Republic: Results of a study. Annual Joint Conference Integrating Technology into Computer Science Education. Proceedings of the 11th annual SIGCSE conference on Innovation and Technology in computer science education. Bologna, Italy, pp. 222–226.

Giedd, J. N., Clasen, L. S., Lenroot, R., Greenstein, D., Wallace, G. L., Ordaz, S. (2006). Puberty-related influences on brain development. *Molecular and Cellular Endocrinology, 254/255*, 154–162.

Gilmore, J. H., Lin, W., Prastawa, M. W., Looney, C. B., Vetsa, Y. S. K., Knickmeyer, R. C., et al. (2007). Regional gray matter growth, sexual dimorphism, and cerebral asymmetry in the neonatal brain. *Journal of Neuroscience, 27*(6), 1255–1260.

Gisborne, T. (1797). *An enquiry into the duties of the female sex*. London: Printed for T. Cadell and W. Davies in the Strand.

Gitau, R., Adams, D., Fisk, N. M., & Glover, V. (2005). Fetal plasma testosterone correlates positively with cortisol. *Archives of Disease in Childhood, Fetal and Neonatal Edition, 90*, F166–F169.

Gladwell, M. (2008). *Outliers: The story of success*. Camberwell, Victoria, Australia: The Penguin Group.

Glick, P., & Fiske, S. (2007). Sex discrimination: The psychological approach. In F. Crosby, M. Stockdale, & S. Ropp (eds.), *Sex discrimination in the workplace: Multidisciplinary perspectives* (pp. 155–187). Malden, MA: Blackwell Publishing.

Golombok, S., & Rust, J. (1993). The pre-school activities inventory: A standardised assessment of gender role in children. *Psychological Assessment, 5*(2), 131–136.

Gonzalez, A. Q., & Koestner, R. (2005). Parental preference for sex of newborn as reflected in positive affect in birth announcements. *Sex Roles, 52*(5/6), 407–411.

Good, C., Aronson, J., & Harder, J. A. (2008). Problems in the pipeline: Stereotype threat and women's achievement in high-level math courses. *Journal of Applied Developmental Psychology, 29*(1), 17–28.

Good, C., Aronson, J., & Inzlicht, M. (2003). Improving adolescents' standardized test performance: An intervention to reduce the effects of stereotype threat. *Journal of Applied Developmental Psychology, 24*(6), 645–662.

Good, C., Rattan, A., & Dweck, C. (unpublished manuscript). Why do women opt out? Sense of belonging and women's representation in mathematics.

Gooden, A. M., & Gooden, M. A. (2001). Gender representation in notable children's picture books: 1995–1999. *Sex Roles, 45*(1/2), 89–101.

Gooren, L. (2006). The biology of human psychosexual differentiation. *Hormones and Behavior, 50,* 589–601.

Gorman, E. H., & Kmec, J. A. (2007). We (have to) try harder: gender and required work effort in Britain and the United States. *Gender and Society, 21*(6), 828–856.

Götz, M. (2008). *Girls and boys and television: A few reminders for more gender sensitivity in children's TV.* Germany: Internationales Zentralinstitut für das Jugend – und Bildungsfernsehen / International Central Institute for Youth and Educational Television (IZI).

Gould, S. (1981). *The mismeasure of man.* London: Penguin Books.

Graham, T., & Ickes, W. (1997). When women's intuition isn't greater than men's. In W. Ickes (ed.), *Empathic accuracy* (pp. 117–143). New York and London: The Guilford Press.

Gray, J. (2008). *Why Mars and Venus collide.* London: Harper Collins.

Green, V. A., Bigler, R., & Catherwood, D. (2004). The variability and flexibility of gender-typed toy play: A close look at children's behavioral responses to counterstereotypic models. *Sex Roles, 51*(7/8), 371–386.

Greenwald, A. G., Poehlman, T. A., Uhlmann, E. L., & Banaji, M. R. (2009). Understanding and using the Implicit Association Test: III. Meta-analysis of predictive validity. *Journal of Personality and Social Psychology, 97*(1), 17–41.

Grimshaw, G. M., Sitarenios, G., & Finegan, J. A. K. (1995). Mental rotation at 7 years – Relations with prenatal testosterone levels and spatial play experiences. *Brain and Cognition, 29*(1), 85–100.

Grön, G., Wunderlich, A. P., Spitzer, M., Tomczak, R., & Riepe, M. W. (2000). Brain activation during human navigation: Gender-different neural networks as substrate of performance. *Nature Neuroscience, 3*(4), 404–408.

Grossi, G. (2008). Science or belief? Bias in sex differences research. In C. Badaloni, A. Drace, O. Gia, C. Levorato, & F. Vidotto (eds.), *Under-representation of women in science and technology* (pp. 93–106). Padova, Italy: Cleup.

Guimond, S. (2008). Psychological similarities and differences between women and men across cultures. *Social and Personality Psychology Compass, 2*(1), 494–510.

Guiso, L., Monte, F., Sapienza, P., & Zingales, L. (2008). Culture, gender, and math. *Science, 320*(5880), 1164–1165.

Gupta, V. K., & Bhawe, N. M. (2007). The influence of proactive personality and stereotype threat on women's entrepreneurial intentions. *Journal of Leadership and Organizational Studies, 13*(4), 73–85.

Gur, R. C., Alsop, D., Glahn, D., Petty, R., Swanson, C. L., Maldjian, J. A., et al. (2000). An fMRI study of sex differences in regional activation to a verbal and a spatial task. *Brain and Language, 74*(2), 157–170.

Gur, R. C., & Gur, R. E. (2007). Neural substrates for sex differences in cognition. In *Why aren't more women in science? Top researchers debate the evidence* (pp. 189–198). Washington, DC: American Psychological Association.

Gur, R. C., Turetsky, B. I., Matsui, M., Yan, M., Bilker, W., Hughett, P., & Gur, R. E. (1999). Sex differences in gray and white brain matter in healthy young adults: Correlations with cognitive performance. *Journal of Neuroscience, 19*(10), 4065–4072.

Gürer, D. W. (2002a). Pioneering women in computer science. *SIGCSE Bulletin, 34*(2), 175–180.

———. (2002b). Women in computing history. *SIGCSE Bulletin, 34*(2), 116–120.

Gurian Institute, Bering, S., & Goldberg, A. (2009). *It's a baby girl! The unique wonder and special nature of your daughter from pregnancy to two years.* San Francisco, CA: Jossey-Bass.

Gurian, M. (2004). *What could he be thinking? A guide to the mysteries of a man's mind.* London: Element.

Gurian, M., & Annis, B. (2008). *Leadership and the sexes: Using gender science to create success in business.* San Francisco, CA: Jossey-Bass.

Gurian, M., Henley, P., & Trueman, T. (2001). *Boys and girls learn differently! A guide for teachers and parents.* San Francisco, CA: Jossey-Bass.

Gurian, M., & Stevens, K. (2004, November). With boys and girls in mind. *Educational Leadership*, 21–26.

———. (2005). *The minds of boys: Saving our sons from falling behind in school and life.* San Francisco, CA: Jossey-Bass.

Gutek, B. A., & Done, R. S. (2001). Sexual harassment. In R. K. Unger (ed.), *Handbook of the psychology of women and gender* (pp. 367–387). Hoboken, NJ: John Wiley & Sons.

Haier, R. J., Jung, R. E., Yeo, R. A., Head, K., & Alkire, M. T. (2005). The neuroanatomy of general intelligence: Sex matters. *NeuroImage, 25*, 320–327.

Haier, R. J., Siegal, B. V., MacLachlan, A., Soderling, E., Lottenberg, S., & Buchsbaum, M. S. (1992). Regional glucose metabolic changes after learning a complex visuospatial/motor task: A positron emission tomographic study. *Brain Research, 570*, 134–143.

Halari, R., Sharma, T., Hines, M., Andrew, C., Simmons, A., & Kumari, V. (2006). Comparable fMRI activity with differential behavioural performance on mental rotation and overt verbal fluency tasks in healthy men and women. *Experimental Brain Research, 169*(1), 1–14.

Hall, G. B. C., Witelson, S. F., Szechtman, H., & Nahmias, C. (2004). Sex differences in functional activation patterns revealed by increased emotion processing demands. *NeuroReport, 15*(2), 219–223.

Hall, J. A. (1978). Gender effects in decoding nonverbal cues. *Psychological Bulletin, 85*(4), 845–857.

———. (1984). *Nonverbal sex differences: Communication accuracy and expressive style.* Baltimore & London: The Johns Hopkins University Press.

Halpern, D. F., Benbow, C. P., Geary, D. C., Gur, R. C., Hyde, J. S., & Gernsbacher, M. A. (2007). The science of sex differences in science and mathematics. *Psychological Science in the Public Interest, 8*(1), 1–51.

Hamilton, C. (2004). *Growth fetish.* London: Pluto Press.

Hamilton, M. C., Anderson, D., Broaddus, M., & Young, K. (2006). Gender stereotyping and under-representation of female characters in 200 popular children's picture books: A twenty-first century update. *Sex Roles, 55*(11/12), 757–765.

Hanlon, H. W., Thatcher, R. W., & Cline, M. J. (1999). Gender differences in the development of EEG coherence in normal children. *Developmental Neuropsychology, 16*(3), 479–506.

Harrington, G. S., & Farias, S. T. (2008). Sex differences in language processing: Functional MRI methodological considerations. *Journal of Magnetic Resonance Imaging, 27*(6), 1221–1228.

Haslam, S. A., & Ryan, M. K. (2008). The road to the glass cliff: Differences in the perceived suitability of men and women for leadership positions in succeeding and failing organizations. *The Leadership Quarterly, 19*(5), 530–546.

Haslanger, S. (2008). Changing the ideology and culture of philosophy: Not by reason (alone). *Hypatia, 23*(2), 210–223.

Hassett, J. M., Siebert, E. R., & Wallen, K. (2008). Sex differences in rhesus monkey toy preferences parallel those of children. *Hormones and Behavior, 54*(3), 359–364.

Hausmann, M., Schoofs, D., Rosenthal, H. E. S., & Jordan, K. (2009). Interactive effects of sex hormones and gender stereotypes on cognitive sex differences – A psychobiological approach. *Psychoneuroendocrinology, 34*, 389–401.

Healy, M. (2006a, May 8). Hunter vs. gatherer (and then some). *Los Angeles Times*, F12.

———. (2006b, July 23). In mental juggling, gender makes a difference. *Telegraph (Nashua, NH)*, Health section.

Heilman, M. E. (2001). Description and prescription: How gender stereotypes prevent women's ascent up the organizational ladder. *Journal of Social Issues, 57*(4), 657–674.

Heilman, M. E., & Chen, J. J. (2005). Same behavior, different consequences: Reactions to men's and women's altruistic citizenship behavior. *Journal of Applied Psychology, 90*(3), 431–441.

Heilman, M. E., Wallen, A. S., Fuchs, D., & Tamkins, M. M. (2004). Penalties for success: Reactions to women who succeed at male gender-typed tasks. *Journal of Applied Psychology, 89*(3), 416–427.

Herman, R. A., Measday, M. A., & Wallen, K. (2003). Sex differences in interest in infants in juvenile rhesus monkeys: Relationship to prenatal androgens. *Hormones and Behavior, 43*, 573–583.

Hersch, J. (2006). Sex discrimination in the labor market. *Foundations and Trends in Microeconomics, 2*(4), 281–361.

Hess, B. B. (1990). Beyond dichotomy: Drawing distinctions and embracing differences. *Sociological Forum, 5*(1), 75–93.

Hewlett, S., Luce, C. B., & Servon, L. J. (2008, December). [Response to letter]. *Harvard Business Review*, 114.

Hewlett, S. A., Luce, C. B., Servon, L. J., Sherbin, L., Shiller, P., Sosnovich, E., & Sumberg, K. (2008). *The Athena factor: Reversing the brain drain in science, engineering, and technology.* Boston: Harvard Business Review.

Hines, M. (2004). *Brain gender.* Oxford and New York: Oxford University Press.

———. (2006a). Do sex differences in cognition cause the shortage of women in science? In S. Ceci & W. Williams (eds.), *Why aren't more women in science? Top researchers debate the evidence* (pp. 101–112). Washington, DC: American Psychological Association.

———. (2006b). Prenatal testosterone and gender-related behaviour. *European Journal of Endocrinology, 155*, S115–S121.

Hines, M., & Alexander, G. (2008). Monkeys, girls, boys and toys: A confirmation. Letter regarding 'Sex differences in toy preferences: Striking parallels between monkeys and humans'. *Hormones and Behavior, 54*, 478–479.

Hines, M., Fane, B. A., Pasterski, V. L., Mathews, G. A., Conway, G. S., & Brook, C. (2003). Spatial abilities following prenatal androgen abnormality: Targeting and mental rotations performance in individuals with congenital adrenal hyperplasia. *Psychoneuroendocrinology, 28*, 1010–1026.

Hines, M., Golombok, S., Rust, J., Johnston, K., Golding, J., and the Avon Longitudinal Study of Parents and Children Study Team (2002). Testosterone during pregnancy and gender role behavior of preschool children: A longitudinal, population study. *Child Development, 73*(6), 1678–1687.

Hinze, S. W. (2004). 'Am I being over-sensitive?' Women's experience of sexual harassment during medical training. *Health: An Interdisciplinary Journal for the Social Study of Health, Illness and Medicine, 8*(1), 101–127.

Hochschild, A. R. (1990). *The second shift*. New York: Avon Books.

Hoff Sommers, C. (2008). Why can't a woman be more like a man? From *The American* online: March/April issue. http://www.american.com/ archive/2008/march-april-magazine-contents/why-can2019t-a-woman-be-more-like-a-man, accessed on March 19, 2008.

Hogg, M. A., & Turner, J. C. (1987). Intergroup behaviour, self-stereotyping and the salience of social categories. *British Journal of Social Psychology, 26*, 325–340.

Hollingworth, L. S. (1914). Variability as related to sex differences in achievement: A critique. *American Journal of Sociology, 21*, 510–530.

Horne, M. (2007, July 29). Gender row cooking up in classroom. *Scotland on Sunday*, 7.

Hornik, R., Risenhoover, N., & Gunnar, M. (1987). The effects of maternal positive, neutral, and negative affective communications on infant responses to new toys. *Child Development, 58*, 937–944.

Houck, M. (2009). Is forensic science a gateway for women in science? *Forensic Science Policy and Management, 1*, 65–69.

Hughes, C., & Cutting, A. L. (1999). Nature, nurture, and individual differences in early understanding of mind. *Psychological Science, 10*(5), 429–432.

Hughes, C., Jaffee, S. R., Happé, F., Taylor, A., Caspi, A., & Moffit, T. E. (2005). Origins of differences in theory of mind: From nature to nurture? *Child Development, 76*(2), 356–370.

Huguet, P., & Régner, I. (2007). Stereotype threat among schoolgirls in quasi-ordinary classroom circumstances. *Journal of Educational Psychology, 99*(3), 545–560.

Hurlbert, A. C., & Ling, Y. (2007). Biological components of sex differences in color preference. *Current Biology, 17*(16), R623–R625.

Hyde, J. S. (2005). The gender similarities hypothesis. *American Psychologist, 60*(6), 581–592.

Hyde, J. S., Lindberg, S. M., Linn, M. C., Ellis, A. B., & Williams, C. C. (2008). Gender similarities characterize math performance. *Science 321*, 494–495.

Hyde, J. S., & Mertz, J. E. (2009). Gender, culture, and mathematics performance. *Proceedings of the National Academy of Sciences, 106*(22), 8801–8807.

Ickes, W. (2003). *Everyday mind reading: Understanding what other people think and feel*. Amherst, NY: Prometheus Books.

Ickes, W., Gesn, P. R., & Graham, T. (2000). Gender differences in empathic accuracy: Differential ability or differential motivation? *Personal Relationships, 7*, 95–109.

Ihnen, S. K. Z., Church, J. A., Petersen, S. E., & Schlaggar, B. L. (2009). Lack of generalizability of sex differences in the fMRI BOLD activity associated with language processing in adults. *NeuroImage, 45*(3), 1020–1032.

Iidaka, T., Okada, T., Murata, T., Omori, M., Kosaka, H., Sadato, N., & Yonekura, Y. (2002). Age-related differences in the medial temporal lobe responses to emotional faces as revealed by fMRI. *Hippocampus, 12*, 352–362.

Im, K., Lee, J.-M., Lyttelton, O., Kim, S. H., Evans, A. C., & Kim, S. I. (2008). Brain size and cortical structure in the adult human brain. *Cerebral Cortex, 18*, 2181–2191.

Inzlicht, M., & Ben-Zeev, T. (2000). A threatening intellectual environment: Why females are susceptible to experiencing problem-solving deficits in the presence of males. *Psychological Science, 11*(5), 365–371.

Itani, J. (1959). Paternal care in the wild Japanese monkey, *Macaca fuscata fuscata. Primates, 2*(1), 61–93.

James, K. (1993). Conceptualizing self with in-group stereotypes: Context and esteem precursors. *Personality and Social Psychology Bulletin, 19*(1), 117–121.

Jeffreys, S. (2008). Keeping women down and out: The strip club boom and the reinforcement of male dominance. *SIGNS: Journal of Women in Culture and Society, 34*(1), 151–173.

Johns, M., Inzlicht, M., & Schmader, T. (2008). Stereotype threat and executive resource depletion: Examining the influence of emotion regulation. *Journal of Experimental Psychology: General, 137*(4), 691–705.

Johnson, W., Carothers, A., & Deary, I. J. (2008). Sex differences in variability in general intelligence. *Perspectives on Psychological Science, 3*(6), 518–531.

Jolls, C. (2002). Is there a glass ceiling? *Harvard Women's Law Journal, 25*, 1–18.

Jordan, K., Wüstenberg, T., Heinze, H.-J., Peters, M., & Jäncke, L. (2002). Women and men exhibit different cortical activation patterns during mental rotation tasks. *Neuropsychologia, 40*(13), 2397–2408.

Josephs, R. A., Newman, M. L., Brown, R. P., & Beer, J. M. (2003). Status, testosterone, and human intellectual performance: Stereotype threat as status concern. *Psychological Science, 14*(2), 158–163.

Jost, J., & Hunyady, O. (2002). The psychology of system justification and the palliative function of ideology. *European Review of Social Psychology, 13*, 111–153.

Jost, J. T., Pelham, B. W., & Carvallo, M. R. (2002). Non-conscious forms of system justification: Implicit and behavioral preferences for higher status groups. *Journal of Experimental Social Psychology, 38*, 586–602.

Jürgensen, M., Hiort, O., Holterhus, P.-M., & Thyen, U. (2007). Gender role behavior in children with XY karyotype and disorders of sex development. *Hormones and Behavior, 51*(3), 443–453.

Kaiser, A., Haller, S., Schmitz, S., & Nitsch, C. (2009). On sex/gender related similarities and differences in fMRI language research. *Brain Research Reviews, 61*(2): 49–59.

Kane, E. W. (2006a). 'No way my boys are going to be like that!': Responses to children's gender nonconformity. *Gender and Society, 20*(149), 149–176.

———. (2006b). *'We put it down in front of him, and he just instinctively knew what to do': Biological determinism in parents' beliefs about the origins of gendered childhoods*. Paper presented at the Annual Meeting of the American Sociological Association, Montreal, Canada.

———. (2008). *Called to account: Parents and children doing gender in everyday interactions*. Paper presented at the Eastern Sociological Society Annual Meeting, New York, NY.

———. (2009). 'I wanted a soul mate:' Gendered anticipation and frameworks of accountability in parents' preferences for sons and daughters. *Symbolic Interaction, 34*(4), 372–389.

Kaplan, G., & Rogers, L. (1994). Race and gender fallacies: The paucity of biological determinist explanations of difference. In E. Tobach & B. Rosoff (eds.), *Challenging racism and sexism: Alternatives to genetic explanations* (pp. 66–92). New York: The Feminist Press.

Kaplan, G., & Rogers, L. (2003). *Gene worship: Moving beyond the nature/nurture debate over genes, brain, and gender*. New York: Other Press.

Kay, A., Wheeler, S., Bargh, J., & Ross, L. (2004). Material priming: The influence of mundane physical objects on situational construal and competitive behavioral choice. *Organizational Behavior and Human Decision Processes, 95*, 83–96.

Kelly, D. J., Liu, S., Ge, L., Quinn, P. C., Slater, A. M., Lee, K., et al. (2007). Cross-race preferences for same-race faces extend beyond the African versus Caucasian contrast in 3-month-old infants. *Infancy, 11*(1), 87–95.

Kemper, T. (1990). *Social structure and testosterone: Explorations of the socio-biosocial chain*. New Brunswick, NJ: Rutgers University Press.

Kiefer, A. K., & Sekaquaptewa, D. (2007). Implicit stereotypes and women's math performance: How implicit gender-math stereotypes influence women's susceptibility to stereotype threat. *Journal of Experimental Social Psychology, 43*(5), 825–832.

Kilbourne, B., England, P., Farkas, G., Beron, K., & Weir, D. (1994). Returns to skill, compensating differentials, and gender bias: Effects of occupational characteristics on the wages of white women and men. *American Journal of Sociology, 100*(3), 689–719.

Killgore, W. D. S., Oki, M., & Yurgelun-Todd, D. A. (2001). Sex-specific developmental changes in amygdala responses to affective faces. *NeuroReport, 12*(2), 427–433.

Killgore, W. D. S., Yurgelun-Todd, D. A. (2001). Sex differences in amygdala activation during the perception of facial affect. *NeuroReport, 12*(11), 2543–2547.

Kimmel, M. S. (2004). *The gendered society* (2nd ed.). New York and Oxford: Oxford University Press.

———. (2008). *The gendered society* (3rd ed.). New York and Oxford: Oxford University Press.

Kimura, D. (1999). *Sex and cognition*. Cambridge, MA: The MIT Press.

———. (2005, February 1). Hysteria trumps academic freedom. *Vancouver Sun*, p. A13.

Kitayama, S., & Cohen, D. (2007). Preface. In S. Kitayama & D. Cohen (eds.), *Handbook of cultural psychology*. New York: The Guilford Press.

Klein, K. J. K., & Hodges, S. D. (2001). Gender differences, motivation, and empathic accuracy: When it pays to understand. *Personality and Social Psychology Bulletin, 27*(6), 720–730.

Knickmeyer, R., Baron-Cohen, S., Fane, B. A., Wheelwright, S., Mathews, G. A., Conway, G. S., et al. (2006). Androgens and autistic traits: A study of individuals with congenital adrenal hyperplasia. *Hormones and Behavior, 50*, 148–153.

Knickmeyer, R., Baron-Cohen, S., Raggatt, P., & Taylor, K. (2005). Foetal testosterone, social relationships, and restricted interests in children. *Journal of Child Psychology & Psychiatry, 46*(2), 198–210.

Knickmeyer, R., Baron-Cohen, S., Raggatt, P., Taylor, K., & Hackett, G. (2006). Fetal testosterone and empathy. *Hormones and Behavior, 49*, 282–292.

Knickmeyer, R. C., Wheelwright, S., Taylor, K., Ragatt, P., Hackett, G., & Baron-Cohen, S. (2005). Gender-typed play and amniotic testosterone. *Developmental Psychology, 41*(3), 517–528.

Koenig, A. M., & Eagly, A. H. (2005). Stereotype threat in men on a test of social sensitivity. *Sex Roles, 52*(7/8), 489–496.

Kolata, G. (1983). Math genius may have hormonal basis. *Science, 222*(4630), 1312.

———. (1995, February 28). Man's world, woman's world? Brain studies point to differences. *New York Times*, C1.

Krendl, A. C., Richeson, J. A., Kelley, W. M., & Heatherton, T. F. (2008). The negative consequences of threat: A functional magnetic resonance imaging investigation of the neural mechanisms underlying women's underperformance in math. *Psychological Science, 19*(2), 168–175.

Kriegeskorte, N., Simmons, W. K., Bellgowan, P. S. F., & Baker, C. I. (2009).

Circular analysis in systems neuroscience: The dangers of double dipping. *Nature Neuroscience, 12*(5), 535–540.

Kunda, Z., & Spencer, S. J. (2003). When do stereotypes come to mind and when do they color judgment? A goal-based theoretical framework for stereotype activation and application. *Psychological Bulletin, 129*(4), 522–544.

Lamb, M. E., Easterbrooks, M. A., & Holden, G. W. (1980). Reinforcement and punishment among preschoolers: Characteristics, effects, and correlates. *Child Development, 51*(4), 1230–1236.

Lamb, M. E., & Roopnarine, J. L. (1979). Peer influences on sex-role development in preschoolers. *Child Development, 50*(4), 1219.

Lamb, S., & Brown, L. (2006). *Packaging girlhood: Rescuing our daughters from marketers' schemes*. New York: St Martin's Press.

Lawrence, P. A. (2006). Men, women, and ghosts in science. *PLoS Biology, 4*(1), 13–15.

Lawson, A. (2007, May 23). The princess gene. *The Age*, 18.

Leaper, C., Anderson, K. J., & Sanders, P. (1998). Moderators of gender effects on parents' talk to their children: A meta-analysis. *Developmental Psychology, 34*(1), 3–27.

Leaper, C., Breed, L., Hoffman, L., & Perlman, C. (2002). Variations in the gender-stereotyped content of children's television cartoons across genres. *Journal of Applied Social Psychology, 32*(8), 1653–1662.

Leeb, R. T., & Rejskind, F. G. (2004). Here's looking at you, kid! A longitudinal study of perceived gender differences in mutual gaze behavior in young infants. *Sex Roles, 50*(1/2), 1–14.

Lehrer, J. (2008, August 17). Of course I love you, and I have the brain scan to prove it – We're looking for too much in brain scans. *Boston Globe*. K1.

Leinbach, M. D., Hort, B. E., & Fagot, B. I. (1997). Bears are for boys: Metaphorical associations in young children's gender stereotypes. *Cognitive Development, 12*, 107–130.

Leinbach, M. D., Hort, B., & Fagot, B. I. (1993). Metaphorical dimensions and the gender-typing of toys. Paper presented at the Symposium conducted at the meeting of the Society for Research in Child Development, New Orleans, Louisiana.

Lenroot, R. K., Gogtay, N., Greenstein, D. K., Wells, E. M., Wallace G. L., Clasen, L. S., et al. (2007). Sexual dimorphism of brain developmental trajectories during childhood and adolescence. *NeuroImage, 36*, 1065–1073.

Leonard, C. M., Towler, S., Welcome, S., Halderman, L. K., Otto, R., Eckert, M. A., & Chiarello, C. (2008). Size matters: Cerebral volume influences sex differences in neuroanatomy. *Cerebral Cortex, 18*(12), 2920–2931.

Levine, S. C., Vasilyeva, M., Lourenco, S. F., Newcombe, N. S., &

Huttenlocher, J. (2005). Socioeconomic status modifies the sex difference in spatial skills. *Psychological Science, 16*(11), 841–845.

Levy, G. D., & Haaf, R. A. (1994). Detection of gender-related categories by 10-month-old infants. *Infant Behavior & Development, 17*(4), 457–459.

Levy, N. (2004). Book review: Understanding blindness. *Phenomenology and the Cognitive Sciences, 3,* 315–324.

Lewontin, R. (2000). *It ain't necessarily so: The dream of the human genome and other illusions.* New York: New York Review of Books.

Liben, L., Bigler, R., & Krogh, H. (2001). Pink and blue collar jobs: Children's judgments of job status and job aspirations in relation to sex of worker. *Journal of Experimental Child Psychology, 79*(4), 346–363.

Lickliter, R., & Honeycutt, H. (2003). Developmental dynamics: Toward a biologically plausible evolutionary psychology. *Psychological Bulletin, 129*(6), 819–835.

Lockwood, P. (2006). 'Someone like me can be successful': Do college students need same-gender role models? *Psychology of Women Quarterly, 30*(1), 36–46.

Lockwood, P., & Kunda, Z. (1997). Superstars and me: Predicting the impact of role models on the self. *Journal of Personality and Social Psychology, 73*(1), 91–103.

Logel, C., Iserman, E. C., Davies, P. G., Quinn, D. M., & Spencer, S. J. (2008). The perils of double consciousness: The role of thought suppression in stereotype threat. *Journal of Experimental Social Psychology, 45*(2), 299–312.

Logel, C., Walton, G. M., Spencer, S. J., Iserman, E. C., von Hippel, W., & Bell, A. E. (2009). Interacting with sexist men triggers social identity threat among female engineers. *Journal of Personality and Social Psychology, 96*(6), 1089–1103.

Luders, E., Steinmetz, H., & Jäncke, L. (2002). Brain size and grey matter volume in the healthy human brain. *NeuroReport, 13*(17), 2371–2374.

Lutchmaya, S., Baron-Cohen, S., & Ragatt, P. (2002). Foetal testosterone and eye contact in 12-month-old human infants. *Infant Behavior and Development, 25,* 327–335.

Lynn, M. (2006, January 12). On bankers and lap dancers. *International Herald Tribune,* 18.

Lytton, H., & Romney, D. M. (1991). Parents' differential socialization of boys and girls: A meta-analysis. *Psychological Bulletin, 109*(2), 267–296.

MacAdam, G. (1914, January 18). Feminist revolutionary principle is biological bosh. *New York Times,* SM2 (online archive).

Machin, S., & Pekkarinen, T. (2008). Global sex differences in test score variability. *Science, 322,* 1331–1332.

Maguire, E. A., Gadian, D. G., Johnsrude, I. S., Good, C. D., Ashburner, J., Frackowiak, R. S. J., et al. (2000). Navigation-related structural change in the hippocampi of taxi drivers. *Proceedings of the National Academy of Sciences, USA, 97*(8), 4398–4403.

Malebranche, N. (1997). *The search after truth* (T. Lennon & P. Olscamp, trans.). Cambridge, UK: Cambridge University Press.

Mareschal, D., Johnson, M. H., Sirois, S., Spratling, M. W., Thomas, M. S. C., & Westermann, G. (2007). *Neuroconstructivism: How the brain constructs cognition* (vol. 1). Oxford: Oxford University Press.

Martin, C. L. (1990). Attitudes and expectations about children with nontraditional and traditional gender roles. *Sex Roles, 22*(3/4), 151–166.

Martin, C. L., Eisenbud, L., & Rose, H. (1995). Children's gender-based reasoning about toys. *Child Development, 66*(5), 1453–1471.

Martin, C. L., & Halverson, C. F. (1981). A schematic processing model of sex typing and stereotyping in children. *Child Development, 52*, 1119–1134.

Martin, C. L., & Ruble, D. (2004). Children's search for gender cues: Cognitive perspectives on gender development. *Current Directions in Psychological Science, 13*(2), 67–70.

Martin, C.L., Ruble, D. N., & Szkrybalo, J. (2002). Cognitive theories of early gender development. *Psychological Bulletin, 128*(6), 903–933.

Martin, K. A. (2005). William wants a doll. Can he have one? Feminists, child care advisors, and gender-neutral child rearing. *Gender & Society, 19*(4), 456–479.

Marton, F., Fensham, P., & Chaiklin, S. (1994). A Nobel's eye view of scientific intuition: Discussions with the Nobel prize–winners in physics, chemistry and medicine (1970–86). *International Journal of Science Education, 16*(4), 457–473.

Marx, D. M., Gilbert, P., Monroe, A., & Cole, C. (unpublished manuscript). Superstars close to me: The effect of role model closeness on performance under threat.

Marx, D. M., & Roman, J. S. (2002). Female role models: Protecting women's math test performance. *Personality and Social Psychology Bulletin, 28*(9), 1183–1193.

Marx, D. M., & Stapel, D. A. (2006a). Distinguishing stereotype threat from priming effects: On the role of the social self and threat-based concerns. *Journal of Personality and Social Psychology, 91*(2), 243–254.

———. (2006b). It's all in the timing: Measuring emotional reactions to stereotype threat before and after taking a test. *European Journal of Social Psychology, 36*, 687–698.

———. (2006c). It depends on your perspective: The role of self-relevance in stereotype-based underperformance. *Journal of Experimental Social Psychology*, *42*, 768–775.

Marx, D. M., Stapel, D. A., & Muller, D. (2005). We can do it: The interplay of construal orientation and social comparisons under threat. *Journal of Personality and Social Psychology*, *88*(3), 432–446.

Mason, M. A., & Goulden, M. (2004). Marriage and baby blues: Redefining gender equity in the academy. *The Annals of the American Academy of Political and Social Science*, *596*(1), 86–103.

Mason, W. A. (2002). The natural history of primate behavioral development: An organismic perspective. In D. J. Lewkowicz & R. Lickliter (eds.), *Conceptions of development: Lessons from the laboratory* (pp. 105–134). New York, London, and Hove, UK: Psychology Press.

Mast, M. S. (2004). Men are hierarchical, women are egalitarian: An implicit gender stereotype. *Swiss Journal of Psychology*, *63*(2), 107–111.

Masters, J. C., Ford, M. E., Arend, R., Grotevant, H. D., & Clark, L. V. (1979). Modeling and labeling as integrated determinants of children's sex-typed imitative behavior. *Child Development*, *50*, 364–371.

Mathews, G. A., Fane, B. A., Conway, G. S., Brook, C.G.D., & Hines, M. (2009). Personality and congenital adrenal hyperplasia: Possible effects of prenatal androgen exposure. *Hormones and Behavior*, *55*, 285–291.

Mathews, G. A., Fane, B. A., Pasterski, V. L., Conway, G. S., Brook, C., & Hines, M. (2004). Androgenic influences on neural asymmetry: Handedness and language lateralization in individuals with congenital adrenal hyperplasia. *Psychoneuroendocrinology*, *29*(6), 810–822.

McCabe, D. P., & Castel, A. D. (2008). Seeing is believing: The effect of brain images on judgments of scientific reasoning. *Cognition*, *107*, 343–352.

McClure, E. B. (2000). A meta-analytic review of sex differences in facial expression processing and their development in infants, children, and adolescents. *Psychological Bulletin*, *126*(3), 424–453.

McCrum, R. (2008, August 24). Up Pompeii with the roguish don. *The Observer*, 22.

McGlone, M. S., & Aronson, J. (2006). Stereotype threat, identity salience, and spatial reasoning. *Journal of Applied Developmental Psychology*, *27*(5), 486–493.

McIntyre, M. H. (2006). The use of digit ratios as markers for perinatal androgen action. *Reproductive Biology and Endocrinology*, *4*(1), 10.

McIntyre, R. B., Lord, C. G., Gresky, D. M., Ten Eyck, L. L., Frye, G. D. J., & Bond, C. F. (2005). A social impact trend in the effects of role models on alleviating women's mathematics stereotype threat. *Current Research in Social Psychology*, *10*(9), 116–136.

McIntyre, R. B., Paulson, R. M., & Lord, C. G. (2003). Alleviating women's mathematics stereotype threat through salience of group achievements. *Journal of Experimental Social Psychology, 39*, 83–90.

Meyer-Bahlburg, H. F., Dolezal, C., Baker, S. W., Carlson, A. D., Obeid, J. S., & New, M. I. (2004). Prenatal androgenization affects gender-related behavior but not gender identity in 5–12-year-old girls with congenital adrenal hyperplasia. *Archives of Sexual Behavior, 33*(2), 97–104.

Meyer-Bahlburg, H. F., Dolezal, C., Zucker, K. J., Kessler, S. J., Schober, J. M., & New, M. I. (2006). The recalled Childhood Gender Questionnaire–revised: A psychometric analysis in a sample of women with congenital adrenal hyperplasia. *Journal of Sex Research, 43*(4), 364–367.

Mill, J. (1869/1988). *The subjection of women*. Indianapolis: Hackett Publishing Company.

Miller, C. F., Lurye, L. E., Zosuls, K. M., & Ruble, D. N. (2009). Accessibility of gender stereotype domains: Developmental and gender differences in children. *Sex Roles, 60*(11/12), 870–881.

Miller, C. F., Trautner, H. M., & Ruble, D. N. (2006). The role of gender stereotypes in children's preferences and behavior. In L. Balter & C. Tamis-LeMonda (eds.), *Child psychology: A handbook of contemporary issues* (2nd ed., pp. 293–323). New York: Psychology Press.

Miller, G. (2008). Growing pains for fMRI. *Science, 320* (5882), 1412–1414.

Moè, A. (2009). Are males always better than females in mental rotation? Exploring a gender belief explanation. *Learning and Individual Differences, 19*(1), 21–27.

Moè, A., & Pazzaglia, F. (2006). Following the instructions! Effects of gender beliefs in mental rotation. *Learning and Individual Differences, 16*(4), 369–377.

Moir, A., & Jessel, D. (1989). *Brain sex: The real difference between men and women*. London: Michael Joseph.

Monastersky, R. (2005, March 4). Primed for numbers. (Lawrence H. Summers). *Chronicle of Higher Education, 51*(26): NA.

Mondschein, E. R., Adolph, K. E., & Tamis-LeMonda, C. S. (2000). Gender bias in mothers' expectations about infant crawling. *Journal of Experimental Child Psychology, 77*(4), 304–316.

Montemurri, P. (2009, April 9). Gloria Steinem: Women's liberation is 'longest revolution'. From *Star-Telegram.com*: http://www.star-telegram.com/living/story/1309400.html. Accessed on November 4, 2009.

Moon, C., Cooper, R. P., & Fifer, W. P. (1993). Two-day-olds prefer their native language. *Infant Behavior and Development, 16*(4), 495–500.

Moore, C. L. (1995). Maternal contributions to mammalian reproductive development and the divergence of males and females. *Advances in the Study of Behavior, 24*, 47–118.

Moore, C. L. (2002). On differences and development. In D. J. Lewkowicz & R. Lickliter (eds.), *Conceptions of development: Lessons from the laboratory* (pp. 57–76). New York: Psychology Press.

Moore, C. L., Dou, H., & Juraska, J. M. (1992). Maternal stimulation affects the number of motor neurons in a sexually dimorphic nucleus of the lumbar spinal cord. *Brain Research, 572*(1/2), 52–56.

Moore, D. S., & Johnson, S. P. (2008). Mental rotation in human infants: A sex difference. *Psychological Science, 19*(11), 1063–1066.

Morantz-Sanchez, R. M. (1985). *Sympathy and science: Women physicians in American medicine.* New York and Oxford: Oxford University Press.

Morgan, L. A., & Martin, K. A. (2006). Taking women professionals out of the office: The case of women in sales. *Gender and Society, 20*(1), 108–128.

Morris, J. (1987). *Conundrum.* Harmondsworth, Middlesex: Penguin Books.

Morris, J. A., Jordan, C. L., & Breedlove, S. M. (2004). Sexual differentiation of the vertebrate nervous system. *Nature Neuroscience, 7*(10), 1034–1039.

Morton, T. A., Haslam, S. A., Postmes, T., & Ryan, M. K. (2006). We value what values us: The appeal of identity-affirming science. *Political Psychology, 27*(6), 823–838.

Morton, T. A., Postmes, T., Haslam, S. A., & Hornsey, M. J. (2009). Theorizing gender in the face of social change: Is there anything essential about essentialism? *Journal of Personality and Social Psychology, 96*(3), 653–664.

Mullarkey, M. J. (2004). Two Harvard women: 1965 to today. *Harvard Journal of Law & Gender, 27*(Spring), 367–379.

Murphy, M. C., Steele, C. M., & Gross, J. J. (2007). Signaling threat: How situational cues affect women in math, science, and engineering settings. *Psychological Science, 18*(10), 879–885.

Mussweiler, T., Rüter, K., & Epstude, K. (2004). The ups and downs of social comparison: Mechanisms of assimilation and contrast. *Journal of Personality and Social Psychology, 87*(6), 832–844.

Nash, A., & Grossi, G. (2007). Picking Barbie's brain: Inherent sex differences in scientific ability? *Journal of Interdisciplinary Feminist Thought, 2*(1), Article 5.

Nash, A., & Krawczyk, R. (1994). *Boys' and girls' rooms revisited: The contents of boys' and girls' rooms in the 1990s.* Paper presented at the Conference on Human Development, Pittsburgh, Pennsylvania.

Neuville, E., & Croizet, J.-C. (2007). Can salience of gender identity impair math performance among 7-8 years old girls? The moderating role of task difficulty. *European Journal of Psychology of Education, XXII*(3), 307–316.

Newcombe, N. S. (2007). Taking science seriously: Straight thinking about spatial sex differences. In S. Ceci & W. Williams (eds.), *Why aren't more women in science? Top researchers debate the evidence* (pp. 69–77). Washington, DC: American Psychological Association.

Newman, M. L., Sellers, J. G., & Josephs, R. A. (2005). Testosterone, cognition, and social status. *Hormones and Behavior, 47*(2), 205–211.

Nguyen, H. H., & Ryan, A. M. (2008). Does stereotype threat affect test performance of minorities and women? A meta-analysis of experimental evidence. *Journal of Applied Psychology, 93*(6), 1314–1334.

Nordenström, A., Servin, A., Bohlin, G., Larsson, A., & Wedell, A. (2002). Sex-typed toy play behavior correlates with the degree of prenatal androgen exposure assessed by CYP21 genotype in girls with congenital adrenal hyperplasia. *Journal of Clinical Endocrinology and Metabolism, 87*(11), 5119–5124.

Norton, M. I., Vandello, J. A., & Darley, J. M. (2004). Casuistry and social category bias. *Journal of Personality and Social Psychology, 87*(6), 817–831.

Nosek, B. A. (2007a). Implicit-explicit relations. *Current Directions in Psychological Science, 16*(2), 65–69.

———. (2007b). Understanding the individual implicitly and explicitly. *International Journal of Psychology, 42*(3), 184–188.

Nosek, B. A., & Hansen, J. (2008). The associations in our heads belong to us: Searching for attitudes and knowledge in implicit evaluation. *Cognition and Emotion, 22*(4), 553–594.

Nosek, B. A., Smyth, F. L., Hansen, J. J., Devos, T., Lindner, N. M., Ranganath, K. A., et al. (2007). Pervasiveness and correlates of implicit attitudes and stereotypes. *European Review of Social Psychology, 18*(1), 36–88.

Nosek, B. A., Smyth, F. L., Sriram, N., Lindner, N. M., Devos, T., Ayala, A., et al. (2009). National differences in gender-science stereotypes predict national sex differences in science and math achievement. *Proceedings of the National Academy of Sciences, 106*(26), 10593–10597.

Novell, C. (2004). *Disney princess: How to be a princess.* Camberwell, Victoria: Penguin Books.

Nowlan, M. (2006, October 2). Women doctors, their ranks growing, transform medicine. *Boston Globe*, C1.

O'Boyle, M. W. (2005). Some current findings on brain characteristics of the mathematically gifted adolescent. *International Education Journal, 6*(2), 247–251.

O'Boyle, M. W., Cunnington, R., Silk, T. J., Vaughn, D., Jackson, G., Syngeniotis, A., & Egan, G. F. (2005). Mathematically gifted male adolescents activate a unique brain network during mental rotation. *Cognitive Brain Research, 25*(2), 583–587.

Oberman, L. M., Hubbard, E. M., McCleery, J. P., Altschuler, E. L., Ramachandran, V. S., & Pineda, J. A. (2005). EEG evidence for mirror neuron dysfunction in autism spectrum disorders. *Cognitive Brain Research, 24*, 190–198.

Ohnishi, T., Moriguchi, Y., Matsuda, H., Mori, T., Hirakata, M., Imabayashi, E., et al. (2004). The neural network for the mirror system and mentalizing in normally developed children: An fMRI study. *NeuroReport, 15*(9), 1483–1487.

Onorato, R. S., & Turner, J. C. (2004). Fluidity in the self-concept: The shift from personal to social identity. *European Journal of Social Psychology, 34*, 257–278.

Orenstein, P. (2000). *Flux: Women on sex, work, love, kids, and life in a half-changed world*. New York: Anchor Books.

Orzhekhovskaia, N. S. (2005). [Sex dimorphism of neuron-glia correlations in the frontal areas of the human brain]. *Morfologiia, 127*(1), 7–9.

Paechter, C. (2007). *Being boys, Being girls: Learning masculinities and femininities*. Maidenhead, Berkshire: Open University Press.

Paoletti, J. B. (1997). The gendering of infants' and toddlers' clothing in America. In K. Martinez & K. L. Ames (eds.), *The material culture of gender: The gender of material culture* (pp. 27–35). Hanover, NH, and London: University Press of New England.

Pasterski, V. L., Geffner, M. E., Brain, C., Hindmarsh, P., Brook, C., & Hines, M. (2005). Prenatal hormones and postnatal socialization by parents as determinants of male-typical toy play in girls with congenital adrenal hyperplasia. *Child Development, 76*(1), 264–278.

Patterson, M. M., & Bigler, R. S. (2006). Preschool children's attention to environmental messages about groups: Social categorization and the origins of intergroup bias. *Child Development, 77*(4), 847–860.

Pease, A., & Pease, B. (2008). *Why men don't listen and women can't read maps*. Pease International Pty Ltd.

Penner, A. (2008). Gender differences in extreme mathematical achievement: An international perspective on biological and social factors. *American Journal of Sociology, 114*, S138–170.

Peplau, L. A., & Fingerhut, A. (2004). The paradox of the lesbian worker. *Journal of Social Issues, 60*(4), 719–735.

Phelan, J. E., Moss-Racusin, C. A., & Rudman, L. A. (2008). Competent yet out in the cold: Shifting criteria for hiring reflect backlash toward agentic women. *Psychology of Women Quarterly, 32*, 406–413.

Pierce, A. (2009, February 23). The epitome of a 'very pretty' Tory lady; Unseen papers show how a young Thatcher charmed BBC executives. *Daily Telegraph*, 5.

Pike, J. J., & Jennings, N. A. (2005). The effects of commercials on children's perceptions of gender appropriate toy use. *Sex Roles, 52*(1/2), 83–91.

Pinker, S. (2005, February 14). Sex Ed. From the *New Republic*: http://www.tnr.com/article/sex-ed. Accessed November 18, 2009.

Pinker, S. (2008). *The sexual paradox: Men, women, and the real gender gap*. New York: Scribner.

Poldrack, R. A. (2006). Can cognitive processes be inferred from neuroimaging data? *Trends in Cognitive Sciences, 10*(2), 59–63.

Poldrack, R. A., & Wagner, A. D. (2004). What can neuroimaging tell us about the mind? Insights from prefrontal cortex. *Current Directions in Psychological Science, 13*(5), 177–181.

Pomerleau, A., Bolduc, D., Malcuit, G., & Cossette, L. (1990). Pink or blue: Environmental gender stereotypes in the first two years of life. *Sex Roles, 22*(5/6), 359–367.

Poulin-Dubois, D., Serbin, L. A., Eichstedt, J. A., Sen, M. G., & Beissel, C. F. (2002). Men don't put on make-up: Toddlers' knowledge of the gender stereotyping of household activities. *Social Development, 11*(2), 166–181.

Prime, J., Jonsen, K., Carter, N., & Maznevski, M. L. (2008). Managers' perceptions of women and men leaders: A cross cultural comparison. *International Journal of Cross Cultural Management, 8*(2), 171–210.

Pronin, E., Steele, C. M., & Ross, L. (2004). Identity bifurcation in response to stereotype threat: Women and mathematics. *Journal of Experimental Social Psychology, 40*, 152–168.

Puts, D. A., McDaniel, M. A., Jordan, C. L., & Breedlove, S. M. (2008). Spatial ability and prenatal androgens: Meta-analyses of congenital adrenal hyperplasia and digit ratio (2D:4D) studies. *Archives of Sexual Behavior, 37*(1), 100–111.

Quinn, P. C., & Liben, L. S. (2008). A sex difference in mental rotation in young infants. *Psychological Science, 19*(11), 1067–1070.

Quinn, P. C., Yahr, J., Kuhn, A., Slater, A. M., & Pascalis, O. (2002). Representation of the gender of human faces by infants: A preference for female. *Perception, 31*(9), 1109–1121.

Racine, E., Bar-Ilan, O., & Illes, J. (2005). fMRI in the public eye. *Nature Reviews Neuroscience, 6*(2), 159–164.

Raingruber, B. J. (2001). Settling into and moving in a climate of care: Styles and patterns of interaction between nurse psychotherapists and clients. *Perspectives in Psychiatric Care, 37*(1), 15–27.

Realo, A., Allik, J., Nõlvak, A., Valk, R., Ruus, T., Schmidt, M., et al. (2003). Mind-reading ability: Beliefs and performance. *Journal of Research in Personality, 37*, 420–445.

Rhode, D. L. (1997). *Speaking of sex: The denial of gender inequality.* Cambridge, MA: Harvard University Press.

Ridgeway, C. L., & Correll, S. J. (2004). Unpacking the gender system: A theoretical perspective on gender beliefs and social relations. *Gender & Society, 18*(4), 510–531.

Rivers, C., & Barnett, R. C. (2007, October 28). The difference myth. *Boston Globe*, F1.

Rochat, P. (2001). *The infant's world.* Cambridge, MA: Harvard University Press.

Rogers, K., Dziobek, I., Hassenstab, K., Wolf, O., & Convit, A. (2007). Who cares? Revisiting empathy in Asperger Syndrome. *Journal of Autism and Developmental Disorders, 37*(4), 709–715.

Rogers, L. (1999). *Sexing the brain.* London: Weidenfeld & Nicolson.

Rogers, L. J., Zucca, P., & Vallortigara, G. (2004). Advantages of having a lateralized brain. *Proceedings of the Royal Society of London, Ser. B, 271*, S420–S422.

Romanes, G. J. (1887/1987). Mental differences between men and women. In D. Spender (ed.), *Education papers: Women's quest for equality in Britain, 1850–1912.* London and New York: Routledge and Kegan Paul.

Rosenblatt, J. S. (1967). Nonhormonal basis of maternal behavior in the rat. *Science, 156*(3781), 1512–1513.

Roth, L. M. (2004). Bringing clients back in: Homophily preferences and inequality on Wall Street. *Sociological Quarterly, 45*(4), 613–635.

Rothman, B. (1988). *The tentative pregnancy: Prenatal diagnosis and the future of motherhood.* London: Pandora.

Ruble, D., Lurye, L., & Zosuls, K. (2008). Pink frilly dresses (PFD) and early gender identity [Electronic Version]. *Princeton Report on Knowledge.* http://www.princeton.edu/prok/issues/2-2/pink_frilly.xml. Accessed on April 23, 2008.

Rudman, L. A. (1998). Self-promotion as a risk factor for women: The costs and benefits of counterstereotypical impression management. *Journal of Personality and Social Psychology, 74*(3), 629–645.

Rudman, L. A., & Glick, P. (1999). Feminized management and backlash toward agentic women: The hidden costs to women of a kinder, gentler image of middle managers. *Journal of Personality and Social Psychology, 77*(5), 1004–1010.

———. (2001). Prescriptive gender stereotypes and backlash toward agentic women. *Journal of Social Issues, 57*(4), 743–762.

———. (2008). *The social psychology of gender: How power and intimacy shape gender relations.* New York: The Guilford Press.

Rudman, L. A., & Heppen, J. B. (2003). Implicit romantic fantasies and women's interest in personal power: A glass slipper effect? *Personality and Social Psychology Bulletin, 29*(11), 1357–1370.

Rudman, L. A., & Kilianski, S. E. (2000). Implicit and explicit attitudes toward female authority. *Personality and Social Psychology Bulletin, 26*(11), 1315–1328.

Rudman, L. A., Phelan, J. E., & Heppen, J. B. (2007). Developmental sources of implicit attitudes. *Personality and Social Psychology Bulletin, 33*(12), 1700–1713.

Rudman, L. A., Phelan, J. E., Moss-Rascusin, C. A., & Nauts, S. (manuscript submitted for publication). Status incongruity and backlash effects: Defending the gender hierarchy motivates prejudice toward female leaders.

Rush, E., & La Nauze, A. (2006). *Corporate paedophilia: Sexualisation of children in Australia*. The Australia Institute, Canberra.

Russett, C. E. (1989). *Sexual science: The Victorian construction of womanhood*. Cambridge, MA: Harvard University Press.

Ryan, M. K., David, B., & Reynolds, K. J. (2004). Who cares? The effect of gender and context on the self and moral reasoning. *Psychology of Women Quarterly, 28*, 246–255.

Ryan, M. K., Haslam, S. A., Hersby, M. D., Kulich, C., & Atkins, C. (2007). Opting out or pushed off the edge? The glass cliff and the precariousness of women's leadership positions. *Social and Personality Psychology Compass, 1*(1), 266–279.

Sandnabba, N. K., & Ahlberg, C. (1999). Parents' attitudes and expectations about children's cross-gender behavior. *Sex Roles, 40*(3/4), 249–263.

Sani, F., Bennett, M., Mullally, S., & MacPherson, J. (2003). On the assumption of fixity in children's stereotypes: A reappraisal. *British Journal of Developmental Psychology, 21*, 113–124.

Sapolsky, R. (1997, October). A gene for nothing. From *Discover Magazine* online: http://discovermagazine.com/1997/oct/agenefornothing1242. Accessed on December 12, 2009.

Sax, L. (2005, January 23). Too few women – figure it out. *Los Angeles Times*, M5.
———. (2006). *Why gender matters: What parents and teachers need to know about the emerging science of sex differences*. New York: Broadway Books.

Schaffer, A. (2008, July 1). The sex difference evangelists. From *Slate*: http://www.slate.com/id/2194486/entry/2194487. Accessed on June 11, 2009.

Schilt, K. (2006). Just one of the guys? How transmen make gender visible at work. *Gender and Society, 20*(4), 465–490.

Schmader, T., & Johns, M. (2003). Converging evidence that stereotype threat reduces working memory capacity. *Journal of Personality and Social Psychology,* *85*(3), 440–452.

Schmader, T., Johns, M., & Barquissau, M. (2004). The costs of accepting gender differences: The role of stereotype endorsement in women's experience in the math domain. *Sex Roles, 50*(11/12), 835–850.

Schmader, T., Johns, M., & Forbes, C. (2008). An integrated process model of stereotype threat effects on performance. *Psychological Review, 115*(2), 336–356.

Schneider, F., Habel, U., Kessler, C., Salloum, J. B., & Posse, S. (2000). Gender differences in regional cerebral activity during sadness. *Human Brain Mapping, 9,* 226–238.

Schweder, R. S., & Sullivan, M. A. (1993). Cultural psychology: Who needs it? *Annual Review of Psychology, 44,* 497–523.

Seger, C. R., Smith, E. R., & Mackie, D. M. (2009). Subtle activation of a social categorization triggers group-level emotions. *Journal of Experimental Social Psychology, 45*(3), 460–467.

Seibt, B., & Förster, J. (2004). Stereotype threat and performance: How self-stereotypes influence processing by inducing regulatory foci. *Journal of Personality and Social Psychology, 87*(1), 38–56.

Selmi, M. (2005). Sex discrimination in the nineties, seventies style: Case studies in the preservation of male workplace norms. *Employee Rights and Employment Policy Journal, 9,* 1–50.

———. (2007). The work-family conflict: An essay on employers, men and responsibility. *University of St. Thomas Law Journal, 4,* 573–598.

Senior, A. (2009, May 29). The pernicious pinkification of little girls. *The Times* [London], 31.

Serbin, L. A., Connor, J. M., Burchardt, C. J., & Citron, C. C. (1979). Effects of peer presence on sex-typing of children's play behavior. *Journal of Experimental Child Psychology, 27,* 303–309.

Serbin, L. A., Poulin-Dubois, D., & Eichstedt, J. A. (2002). Infants' responses to gender-inconsistent events. *Infancy, 3*(4), 531–542.

Servin, A., Bohlin, G., & Berlin, L. (1999). Sex differences in 1-, 3-, and 5-year-olds' toy-choice in a structured play-session. *Scandinavian Journal of Psychology, 40*(1), 43–48.

Servin, A., Bohlin, G., Nordenstrom, A., & Larsson, A. (2003). Prenatal androgens and gender-typed behavior: A study of girls with mild and severe forms of congenital adrenal hyperplasia. *Developmental Psychology, 39*(3), 440–450.

Shah, J. (2003). Automatic for the people: How representations of significant others implicitly affect goal pursuit. *Journal of Personality and Social Psychology, 84*(4), 661–681.

Shapiro, J. R., & Neuberg, S. L. (2007). From stereotype threat to stereotype threats: Implications of a multi-threat framework for causes, moderators, mediators, consequences, and interventions. *Personality and Social Psychology Review, 11*(2), 107–130.

Sharps, M. J., Price, J. L., & Williams, J. K. (1994). Spatial cognition and gender: Instructional and stimulus influences on mental image rotation performance. *Psychology of Women Quarterly, 18*(3), 413–425.

Sheldon, J. (2004). Gender stereotypes in educational software for young children. *Sex Roles, 51*(7/8), 433–444.

Sherwin, B. B. (1988). A comparative analysis of the role of androgens in human male and female sexual behavior: Behavioral specificity, critical thresholds, and sensitivity. *Psychobiology, 16*(4), 416–425.

Shields, S. (1975). Functionalism, Darwinism, and the psychology of women: A study in social myth. *American Psychologist, 30*(7), 739–754.

———. (1982). The variability hypothesis: The history of a biological model of sex differences in intelligence. *SIGNS: Journal of Women in Culture & Society, 7*(4), 769–797.

Silverberg, A. (2006). *Remarks at an Association for Women in Mathematics panel [entitled 'Lawrence Summers: One year later'].* Presented at the Joint Mathematics Meetings, San Antonio, Texas.

Silvers, J. A., & Haidt, J. (2008). Moral elevation can induce nursing. *Emotion, 8*(2), 291–295.

Sinclair, L., & Kunda, Z. (2000). Motivated stereotyping of women: She's fine if she praised me but incompetent if she criticized me. *Personality and Social Psychology Bulletin, 26*(11), 1329–1342.

Sinclair, S., Hardin, C. D., & Lowery, B. S. (2006). Self-stereotyping in the context of multiple social identities. *Journal of Personality and Social Psychology, 90*(4), 529–542.

Sinclair, S., Huntsinger, J., Skorinko, J., & Hardin, C. D. (2005). Social tuning of the self: Consequences for the self-evaluations of stereotype targets. *Journal of Personality and Social Psychology, 89*(2), 160–175.

Sinclair, S., & Lun, J. (2006). Significant other representations activate stereotypic self-views among women. *Self and Identity, 5*, 196–207.

Singer, T., Seymour, B., O'Doherty, J., Kaube, H., Dolan, R. J., & Frith, C. D. (2004). Empathy for pain involves the affective but not sensory components of pain. *Science, 303*(1157), 1157–1162.

Singer, T., Seymour, B., O'Doherty, J. P., Stephan, K. E., Dolan, R. J., & Frith,

C. D. (2006). Empathic neural responses are modulated by the perceived fairness of others. *Nature, 439*(7075), 466–469.

Singh, H., & O'Boyle, M. W. (2004). Interhemispheric interaction during global-local processing in mathematically gifted adolescents, average-ability youth, and college students. *Neuropsychology, 18*(2), 371–377.

Skuse, D. H. (2009). Is autism really a coherent syndrome in boys, or girls? *British Journal of Psychology, 100*, 33–37.

Smith, E. R., & DeCoster, J. (2000). Dual-process models in social and cognitive psychology: Conceptual integration and links to underlying memory systems. *Personality and Social Psychology Review, 4*(2), 108–131.

Smith, J. (1998). *Different for girls: How culture creates women.* London: Vintage.

Smith, K. (2005). Prebirth gender talk: A case study in gender socialization. *Women and Language, 28*(1), 49–54.

Smith, S., & Cook, C. (2008). *Gender stereotypes: An analysis of popular films and TV.* Geena Davis Institute on Gender in Media: http://www.thegeenadavisinstitute.org.

Sommer, I. E. C., Aleman, A., Bouma, A., & Kahn, R. S. (2004). Do women really have more bilateral language representation than men? A meta-analysis of functional imaging studies. *Brain, 127*, 1845–1852.

Sommer, I. E. C., Aleman, A., Somers, M., Boks, M. P., & Kahn, R. S. (2008). Sex differences in handedness, asymmetry of the Planum Temporale and functional language lateralization. *Brain Research, 1206*, 76–88.

Spelke, E. S. (2005). Sex differences in intrinsic aptitude for mathematics and science? A critical review. *American Psychologist, 60*(9), 950–958.

Spelke, E. S., & Grace, A. D. (2006). Abilities, motives, and personal styles. *American Psychologist, 61*(7), 725–726.

Spencer, S. J., Steele, C. M., & Quinn, D. M. (1999). Stereotype threat and women's math performance. *Journal of Experimental Social Psychology, 35*(1), 4–28.

Stangor, C., Carr, C., & Kiang, L. (1998). Activating stereotypes undermines task performance expectations. *Journal of Personality and Social Psychology, 75*(5), 1191–1197.

Steele, C. M. (1997). A threat in the air: How stereotypes shape intellectual identity and performance. *American Psychologist, 52*(6), 613–629.

Steele, J. R., & Ambady, N. (2006). 'Math is hard!' The effect of gender priming on women's attitudes. *Journal of Experimental Social Psychology, 42*(4), 428–436.

Steinpreis, R. E., Anders, K. A., & Ritzke, D. (1999). The impact of gender on the review of the curricula vitae of job applicants and tenure candidates: A national empirical study. *Sex Roles, 41*(7/8), 509–528.

Stone, P. (2007). *Opting out? Why women really quit careers and head home.* Berkeley: University of California Press.

Strack, F., & Deutsch, R. (2004). Reflective and impulsive determinants of social behavior. *Personality and Social Psychology Review, 8*(3), 220–247.

Summers, L. (2005). Remarks at National Bureau of Economics Research conference on diversifying the science & engineering workforce, Cambridge, Massachusetts. Available at http://www.president.harvard.edu/speeches/summers_2005/nber.php. Accessed January 7, 2009.

Tajfel, H., & Turner, J. C. (1986). The social identity theory of intergroup behavior. In S. Worchel & W. Austin (eds.), *Psychology of intergroup relations* (pp. 7–24). Chicago: Nelson-Hall Publishers.

Tavris, C. (1992). *The mismeasure of woman: Why women are not the better sex, the inferior sex, or the opposite sex.* New York: Touchstone.

Telford, L. (2003). *My pretty princess beauty kit and book.* New York: Disney Press.

Tenenbaum, H. R., & Leaper, C. (2002). Are parents' gender schemas related to their children's gender-related cognitions? A meta-analysis. *Developmental Psychology, 38*(4), 615–630.

Tepper, C. A., & Cassidy, K. W. (1999). Gender differences in emotional language in children's picture books. *Sex Roles, 40*(3/4), 265–280.

Thoman, D. B., White, P. H., Yamawaki, N., & Koishi, H. (2008). Variations of gender-math stereotype content affect women's vulnerability to stereotype threat. *Sex Roles, 58*, 702–712.

Thomas, G., & Maio, G. R. (2008). Man, I feel like a woman: When and how gender-role motivation helps mind-reading. *Journal of Personality and Social Psychology, 95*(5), 1165–1179.

Thompson, S. K. (1975). Gender labels and early sex role development. *Child Development, 46*, 339–347.

Thompson, T. L., & Zerbinos, E. (1995). Gender roles in animated cartoons: Has the picture changed in 20 years? *Sex Roles, 32*(9/10), 651–673.

Tichenor, V. (2005). Maintaining men's dominance: Negotiating identity and power when she earns more. *Sex Roles, 53*(3/4), 191–205.

Trautner, H. M., Ruble, D. N., Cyphers, L., Kirsten, B., Behrendt, R., & Hartmann, P. (2005). Rigidity and flexibility of gender stereotypes in childhood: Developmental or differential? *Infant and Child Development, 14*(4), 365–381.

Trecker, J. (1974). Sex, science and education. *American Quarterly, 26*(4), 352–366.

Turner-Bowker, D. M. (1996). Gender stereotyped descriptors in children's picture books: Does 'Curious Jane' exist in the literature? *Sex Roles, 35*(7/8), 461–488.

Uddin, L. Q., Kaplan, J. T., Molnar-Szakacs, I., Zaidel, E., & Iacoboni, M. (2005). Self-face recognition activates a frontoparietal 'mirror' network in the right hemisphere: An event-related fMRI study. *NeuroImage, 25*, 926–935.

Udry, J. R. (2000). Biological limits of gender construction. *American Sociological Review, 65*, 443–457.

Uhlmann, E. L., & Cohen, G. L. (2005). Constructed criteria. *Psychological Science, 16*(6), 474–480.

University of Pennsylvania Medical Center (1999). Sex differences found in proportions of gray and white matter in the brain. http://www.sciencedaily. com/releases/1999/05/990518072823.htm. Accessed November 6, 2008.

Unterrainer, J., Wranek, U., Staffen, W., Gruber, T., & Ladurner, G. (2000). Lateralized cognitive visuospatial processing: Is it primarily gender-related or due to quality of performance? *Neuropsychobiology, 41*(2), 95–101.

Valian, V. (1998). *Why so slow? The advancement of women.* Cambridge, MA, and London: MIT Press.

van Anders, S. M., & Watson, N. V. (2006). Social neuroendocrinology: Effects of social contexts and behaviors on sex steroids in humans. *Human Nature, 17*(2), 212–237.

van de Beek, C., Thijssen, J. H. H., Cohen-Kettenis, P. T., van Goozen, S. H. M., & Buitelaar, J. K. (2004). Relationships between sex hormones assessed in amniotic fluid, and maternal and umbilical cord serum: What is the best source of information to investigate the effects of fetal hormonal exposure? *Hormones and Behavior, 46*(5), 663–669.

van de Beek, C., van Goozen, S. H. M., Buitelaar, J. K., & Cohen-Kettenis, P. T. (2009). Prenatal sex hormones (maternal and amniotic fluid) and gender-related play behavior in 13-month-old infants. *Archives of Sexual Behavior, 38*, 6–15.

Verghis, S. (2009, May 2). Triumph of the few. *Good Weekend,* 21–26.

Vespa, J. (2009). Gender ideology construction: A life course and intersectional approach. *Gender and Society, 23*(3), 363–387.

Voracek, M., & Dressler, S. G. (2006). Lack of correlation between digit ratio (2D:4D) and Baron-Cohen's 'Reading the Mind in the Eyes' test, empathy, systemising, and autism-spectrum quotients in a general population sample. *Personality and Individual Differences, 41*, 1481–1491.

Voyer, D., Voyer, S., & Bryden, M. P. (1995). Magnitude of sex differences in spatial abilities: A meta-analysis and consideration of critical variables. *Psychological Bulletin, 117*(2), 250–270.

Vul, E., Harris, C., Winkelman, P., & Pashler, H. (2009). Puzzlingly high correlations in fMRI studies of emotion, personality, and social cognition. *Perspective on Psychological Science, 4*(3), 274–290.

Wager, T. D., Phan, K. L., Liberzon, I., & Taylor, S. F. (2003). Valence, gender, and lateralization of functional brain anatomy in emotion: A meta-analysis of findings from neuroimaging. *NeuroImage, 19*(3), 513–531.

Walker, K. (2008, August 13). All pink and sparkly [Letter to the Editor]. *The Age*, 12.

Wallen, K. (1996). Nature needs nurture: The interaction of hormonal and social influences on the development of behavioral sex differences in rhesus monkeys. *Hormones and Behavior, 30*(4), 364–378.

———. (2005). Hormonal influences on sexually differentiated behavior in nonhuman primates. *Frontiers in Neuroendocrinology, 26*, 7–26.

Wallentin, M. (2009). Putative sex differences in verbal abilities and language cortex: A critical review. *Brain and Language, 108*(3), 175–183.

Walton, G. M., & Spencer, S. J. (2009). Latent ability: grades and test scores systematically underestimate the intellectual ability of negatively stereotyped students. *Psychological Science, 20*(9): 1132–1139.

Weichselbaumer, D., & Winter-Ebmer, R. (2005). A meta-analysis of the international gender wage gap. *Journal of Economic Surveys, 19*(3), 479–511.

Weil, E. (2008, March 2). Teaching to the testosterone. *New York Times Magazine*, 38.

Weisberg, D. S. (2008). Caveat lector: The presentation of neuroscience information in the popular media. *Science Review of Mental Health Practice, 6*(1), 51–56.

Weisberg, D. S., Keil, F. C., Goodstein, J., Rawson, E., & Gray, J. R. (2008). The seductive allure of neuroscience explanations. *Journal of Cognitive Neuroscience, 20*(3), 470–477.

Weitzman, L. J., Eifler, D., Hokada, E., & Ross, C. (1972). Sex-role socialization in picture books for preschool children. *American Journal of Sociology, 77*(6), 1125–1150.

Weitzman, N., Birns, B., & Friend, R. (1985). Traditional and nontraditional mothers' communication with their daughters and sons. *Child Development, 56*(4), 894–898.

Welnsteln, N., Przybylski, A., & Ryan, R. (2009). Can nature make us more caring? Effects of immersion in nature on intrinsic aspirations and generosity. *Personality and Social Psychology Bulletin, 35*(10), 1315–1329.

Westermann, G., Mareschal, D., Johnson, M. H., Sirois, S., Spratling, M. W., & Thomas, M. S. C. (2007). Neuroconstructivism. *Developmental Science, 10*(1), 75–83.

Wexler, B. (2006). *Brain and culture: Neurobiology, ideology, and social change.* Cambridge, MA: The MIT Press.

Wheeler, S., DeMarree, K., & Petty, R. E. (2007). Understanding the role of the self in prime-to-behavior effects: The active-self account. *Personality and Social Psychology Review 11*(3), 234–261.

Williams, C. L. (1992). The glass escalator: Hidden advantages for men in the 'female' professions. *Social Problems, 39*(3), 253–266.

Wingfield, A. (2009). Racializing the glass escalator: Reconsidering men's experiences with women's work. *Gender and Society, 23*(1), 5–26.

Woodzicka, J. A., & LaFrance, M. (2001). Real versus imagined gender harassment. *Journal of Social Issues, 57*(1), 15–30.

Wraga, M., Helt, M., Jacobs, E., & Sullivan, K. (2006). Neural basis of stereotype-induced shifts in women's mental rotation performance. *Social Cognitive and Affective Neuroscience, 2*, 12–19.

Wynne-Edwards, K. E. (2001). Hormonal changes in mammalian fathers. *Hormones and Behavior, 40*(2), 139–145.

Wynne-Edwards, K. E., & Reburn, C. J. (2000). Behavioral endocrinology of mammalian fatherhood. *Trends in Ecology & Evolution, 15*(11), 464–468.

Young, R. M., & Balaban, E. (2006). Psychoneuroindoctrinology. *Nature, 443*, 634.

Zahn-Waxler, C., Klimes-Dougan, B., & Slattery, M. J. (2000). Internalizing problems of childhood and adolescence: Prospects, pitfalls, and progress in understanding the development of anxiety and depression. *Development and Psychopathology, 12*, 443–466.

Zosuls, K. M., Ruble, D. N., Tamis-LeMonda, C. S., Shrout, P. E., Bornstein, M. H., & Greulich, F. K. (2009). The acquisition of gender labels in infancy: Implications for sex-typed play. *Developmental Psychology, 45*(3), 688–701.

INDEX

Page numbers beginning with 245 refer to notes.